Space, Earth and Communication

Space, Earth and Communication

Edward W. Ploman

Quorum Books
Westport, Connecticut

Library of Congress Cataloging in Publication Data

Ploman, Edward W., 1926-
    Space, earth, and communication.

    Bibliography: p.
    Includes index.
    1. Artificial satellites in telecommunication.
I. Title
HE9719.P56   1984   384.5'1   84-12969
ISBN 0-89930-094-4 (lib. bdg.)

Library of Congress Catalog Card Number: 84-12969

ISBN: 0-89930-094-4

First published in 1984 by Quorum Books

Greenwood Press
A division of Congressional Information Service, Inc.
88 Post Road West, Westport, Connecticut 06881

Printed in the United States of America

10 9 8 7 6 5 4 3 2 1

Contents

Introduction

It is easy to be seduced by the great satellite game: new openings and vistas, new ways of linking people and places, new money and power. But if we look more closely, communications satellites seem to raise more questions than we can answer, questions both of opportunity and of risk. Communications satellites, located at the intersection of outer space activity and the current changes in our information environment, therefore represent a double set of ambiguities.

The ambiguities of human activity in outer space are obvious. The new opportunities represent a major theme of this book: increase in knowledge, new means to monitor our environment, to assist the navigation of ships and aircraft, to send and receive words, pictures and data in the first truly world-wide network of communications. At the same time, the move into space has increased competition on earth; the increasing militarization of outer space has raised risk and vulnerability to a level which has rightly been branded intolerable.

Seen from space, Earth is a planet, not a motley collection of countries. The very expression 'outer space' implies a global, or rather a planetary perspective. Thus, our traditionally-defined terrestrial environment has been extended so that our immediate neighbourhood now encompasses the solar system. This larger environment also points to the fundamental unity of mankind, and of all human beings as the inhabitants of a small planet in a large universe. In this perspective there is no excuse for not putting our terrestrial house in order. The key problems are clear-cut and, in principle, so are the solutions: to safeguard outer space for peaceful uses and to ensure that all human beings share in the wealth not only of Earth but also of space.

If we look at satellites from the other angle, from the perspective of communications and information, we face a different situation. Perhaps this situation is less dangerous but more insidious and subtle, representing another level of complexity: there are few clear-cut problems and we do not seem able even to ask the right questions.

Expressions such as 'the communications revolution' and 'the information age' are used to encompass processes and phenomena which we vaguely but profoundly feel are changing society and our ways of looking at society, our relations with work, leisure and learning, our relations with each other, and perhaps with ourselves. These expressions also indicate that we have changed our perspective: it is only in recent years that, for the first time, communications and information have themselves become issues in society.

So we face both new questions and old questions in new guises. Reliable and fast communication through satellites quickens the pace and increases the density of contacts world-wide without any corresponding change in attitudes. In the enthusiastic acceptance of new communications technology, concentration on technical factors and short-term, narrow cost-benefit analyses has resulted in the long-term social, cultural and individual aspects being left out. When these issues are addressed at all, it has generally been after the technology has been developed and made economically feasible, and therefore after it has attracted powerful interests and found a self-justifying momentum of its own.

This book, then, is an attempt to put the advent of satellite communications in the context both of the space age and the information age. The context has many dimensions: political, economic, military. It includes education, development and relations between countries. It stretches from the most advanced technologies to the development of international law. The many issues involved are also demonstrated by the amount of pertinent information: torrents of articles, data by the million, entire rooms full of committee reports. It is a difficult task to trace a path through what could easily become a case of information overload.

The book has been structured so that the first two chapters, on the advent of the space age and the communications revolution, set the stage and give the background for a description of satellite technology and systems. The great satellite game, the struggle for 'satellite power' and the difficulties of coping with this new technology are described in three chapters, interwoven with a discussion of the uses and users of communications satellites and of the problems of formulating policy and law. Finally, an attempt has been made to put satellite communication in the context of changes in international relations, in practice and in analysis, in the light of three main concerns: vulnerability, security and interdependence, and to draw some conclusions in relation to an agenda for the future.

1 The space age: a new dimension

One picture tells more than ten thousand words. The ancient Chinese proverb has never been so well demonstrated as when the first picture of earth was transmitted from outer space. This image stands as one of the most signifcant symbols of our age, at the intersection of developments which define our future.

Rocketry and information technology combined to produce the image of earth from outer space. To achieve this photograph we had to develop the capacity to overcome earth's gravity; to build, launch and control a space vehicle and to equip this spacecraft with communication instruments that could capture and transmit the picture to earth. Earth depicted in a photograph from outer space is thus as much an expression of the space age as of the new information age in which communication satellites play such a crucial role.

Antecedents

Our first tentative exploration and use of outer space represents the convergence of visions from many cultures and draws upon centuries of scientific and technical developments. The idea of spaceflight is anything but new. We find it in the myths and legends of all parts of the world: gods have always moved freely in the heavens. Even though the Greek hero Icarus perished in his attempt to reach the sun, the sophist Lucian of Samosata, already in AD 160, described an imaginary voyage to the moon. There seems then to have been a moratorium on imaginary space travel until, in the Renaissance, the earth was dethroned from its place at the centre of the universe. The moon again exercised an irresistible attraction for the imagination of such writers as Cyrano de Bergerac. Later, many others—Voltaire, Dumas, Jules Verne, Edgar Allan Poe and H. G. Wells—wrote tales of space voyages, to the moon and beyond, often using extravagant means of propulsion. Also, in the infancy of cinema, Méliès made people from the turn of the century visit the man in the moon, a charming and innocent fore-runner to *2001* and *Star Wars*. Voyages into outer space have, of course, been a mainstay of science fiction but the authors have had to imagine further reaches and other universes as man actually landed on the moon and the satellite probes showed us the features of the bodies in our immediate celestial environment.

However, the first condition for any form of space activity has little to do with technological development but concerns the scientific calculation of the speed required to overcome the gravitational pull of the earth. This knowledge could be provided by a very ancient branch of science, celestial mecha-nics, which dates back to the time when man first began to measure the movements of the stars. Celestial mechanics is so old in fact that it has been described as having been put away in mothballs until the new technology of rocketry again gave it a prominent position.

Even the means for undertaking journeys into outer space have distant ancestors, and are the result of many cultures and minds, and wars. The direct forerunners to the launch-ing rockets of today were the Chinese 'arrows of flying fire' which were used against the Mongols in 1232. The rocket continued its military career in Europe during the fourteenth and fifteenth centuries. A more modern version was developed by the Englishman Congreve and used during the Napoleonic wars for bombardments against Boulogne, Copenhagen and Danzig.

Decisive progress in rocket technology was not made until the end of the nineteenth century. The Russian schoolteacher Konstantin Tsiolkovski proved the feasibility of spaceflight through the use of multi-step rockets and published in 1903 the first article on liquid-fuelled rockets. In the United States, modern rocket technology began with Dr Robert H. Goddard who published in 1919 a famous monograph on 'A method of reaching extreme altitudes', and in 1927 launched the first, though short-lived rocket, with liquid fuel. In the mean-time, the German Hermann Oberth was writing his classical works on 'The rockets to the interplanetary spaces' (*Die*

Rakete zu den Planetenräumen) and 'The roads to space travel' (*Die Wege zum Raumschiffart*).

These three names are significant for what was to come later. A direct result of Oberth's work was the creation in Germany of a society for space travel. A similar organization was set up in the United States, while in the USSR the earlier scientific organizations were reorganized in 1934 into a state-supported rocket research institute. Forbidden by the Versailles Treaty to build artillery, the German military in 1933 gave one member of the rocket travel society, the then twenty-two year-old Wernher von Braun, the task of constructing a new experimental military rocket. The results were the rocket base at Peenemünde and the four thousand V2-rockets later dropped over London. In 1945 the race for Berlin was matched by the race for Peenemünde, which was taken over by the Russians together with a great number of scientists, engineers and technicians—but not before the Americans had been able to get hold of von Braun, another group of scientists and some hundred V2-rockets. And this incident set the stage for outer space exploration: a race into outer space which in principle has been reduced to two competitors, although a few other countries have manufactured rockets of various kinds and established rocket bases of their own.

Like the atomic age, the space age has scientific parents but military midwives. With slight rearrangement, the rocket and guidance systems that could deliver a nuclear warhead across an ocean, could also accurately place a vehicle of thousands of kilogrammes in an orbit around the earth, or direct it to the moon. Thus, in the post-war production of intercontinental ballistic missiles there were several types that, in somewhat different versions, could be used for the launching of satellites and other spacecraft.

Advent of the space age

By 1955, both the USSR and the United States were preparing for the launching of satellites under the auspices of the International Geophysical Year, which lasted from July 1957 to the end of 1958 and included the co-operation of scientists from all over the world for the study of earth and sun, and their interrelationship.

One of the most sensational news items of this century states, in a dry and succinct form, that for several years research and experimental work had taken place in the Soviet Union for the manufacturing of artificial satellites and that the first satellite was successfully launched in the USSR on 4 October 1957. The style of this TASS message which announced the launch of the first Sputnik and the advent of the space age then deteriorates. As literature it is not much more colourful than the later exchanges between the American moon landers and mission control.

The 'bleep bleep' of Sputnik was found more interesting than the victory pronouncements by Kruschev. These were quite understandable, however, as were the wide smiles of those Russian experts who at that very moment were participating in the Eighth Conference of the International Astronautical Federation in Barcelona, or those in Washington who were discussing rockets and satellite launches with their American opposite numbers.

The reaction to Sputnik 1 has been described as 'a strange mixture of awe, admiration and fear, the last enhanced, of course, because there had been no warning' (Hynek, quoted in Hayes: 1970, p. 18). Others have talked of the shock wave that hit the United States. Within the United States:

a chain of events was triggered by Sputnik. The post of Special Assistant to the President for Science and Technology (later the Office of Science Adviser) was announced. Vast increases followed, in the missile and then the space programs. Both the House and the Senate established space and science committees. The National Aeronautics and Space Administration (NASA) was created in mid-1958. Military space programs achieved a degreee of respectability within the Pentagon. The State Department established a special office to deal with the international political implications of space. Meanwhile, the United States Information Agency had a difficult time attempting to offset the decline of American stature abroad in the face of the spectacular Russian successes. [Goldsen, 1963: p. 8.]

American prestige suffered a further setback through the failure of the first Vanguard launch in 1957. And even before the Americans had launched their first satellite, Explorer 1, the Russians had managed both Sputnik 2 and the space dog Lajka. However, Explorer hit another first by discovering a radiation belt around the earth, later named after the American scientist van Allen. Gradually, the American technical programme gained momentum and the United States recouped some of its lost reputation with the achievements of 1961 and 1962. By the end of 1962 the United States had successfully launched 120 satellites into earth orbit, including the first active satellite exclusively designed for communications purposes, and six probes into deep space. The Soviet Union had then launched thirty-three satellites into earth orbit and four into deep space. Thus, from the beginning, space activities came under the sign of cold war rivalry. The space race became a big new power game—against the background of a constantly growing arsenal of ballistic missiles.

The capacity to provide space transportation services for a wide variety of spacecraft developed rapidly. The early efforts of the 1950s involved modest requirements and were generally concerned with the launching of relatively simple research experiments. The space programmes of the early 1960s included the launch of prototype communications and meteorological spacecraft, as well as lunar and interplanetary probes.

Within the period of one generation, these advances had led from the launch of the first artificial satellite in 1957 to manned spaceflight, men and robot vehicles on the moon, automatic landers on Mars and Venus, missions past Jupiter and Saturn, to the first partially re-usable launch vehicle and space stations in Earth orbit. . . . We have witnessed the establishment and now routine operation of space communication systems, both international and domestic, of space broadcasting systems, of a global observation system for meteorology, of operational navigation and maritime communication systems and of quasi-operational remote-sensing systems. [UN Document A/CONF. 101/10, pp. 1–2.]

Impact of outer space activities

The year 1982 marked the twenty-fifth anniversary of the advent of the space age. It is therefore fitting to ask: What has been the impact of man's first ventures into outer space? What are the issues? And even more basic: why go into space at all?

There have always been two extreme attitudes to the subject of outer space. There are the enthusiasts, the space buffs for whom outer space represents the only remaining, necessary and totally engrossing new frontier which incidentally will also assist in solving our problems on earth. 'Like all frontiers, space has produced unexpected treasures, generated strong enthusiasts, spawned wild speculations and been enshrouded in myth and false promise' (Deudney, 1982, p. 5). Therefore it should not surprise us that there are those who totally oppose any outer space activity, whether they regard humans in space as a blasphemy, or violently object to the idea of spending human and natural resources in space rather than on pressing terrestrial problems. There is of course more to outer space than expressed by either of these extreme approaches.

In many respects fundamental attitudes towards space are philosophical in kind, even though they are often not expressed in those terms. There are many who feel that, because spaceflight is possible, it is in human nature to feel that it should be achieved. The frontier, new horizons, have always been a challenge to some individuals and form part of old and, in some societies, modern mythology. There are those who feel that humanity is destined to step beyond its earthly bounds, just as our distant ancestors once crawled out of the sea. The ultimate dream is one of man or his descendants spreading to new habitats when earth is no longer habitable—a dream of human survival on a cosmic scale. Seen from another angle,

the novelty will be rather that men will have outgrown at last the diseases of mere infancy, and will be able to enter, for the first time, into the more dangerous, more troubled, but ampler and richer and more conscious life of adolescence. [Stapledon, in Clarke, 1970: p. 268.]

In this vein, some think that

the idea of travelling to other celestial bodies reflects to the highest degree the independence and agility of the human mind. It lends ultimate dignity to man's technical and scientific endeavours. Above all, it touches on the philosophy of his very existence. [Ehricke, in Clarke, 1970: p. 285.]

But even those who will not admit to philosophical reflections have been influenced by space activities in many ways. Images of space, real and fictional, have penetrated our media and our awareness. They are reflected in the renewed interest in astronomy and in the move by science fiction in our culture from slum-dwelling to the mainstream. Images from space have also changed perceptions in a more directly relevant manner.

As early as 1948, the astronomer Fred Hoyle predicted that 'once a photograph of earth, taken from outside is available—once the sheer isolation of earth becomes plain, a new idea as powerful as any in history will be let loose' (Hoyle 1971, p. 129). This idea is the recognition of 'one world', of our planet as a unique and precious place, of spaceship earth which will combine a space image with a global approach to our environment. Buckminster Fuller's 'operating manual for spaceship earth' concerns the management not of nations, nor of regions but of our planet and equals the attempts to evolve legal bases for securing the integrity of the earth-space environment. Outer space has given us a new sense of the place of earth and of humanity in the cosmic order. And more, we have been given a widening environment. Our immediate neighbourhood has expanded to become no less than the solar system. If, as seems to be the case, there is no other intelligent life form in this small chamber of a large universe, human responsibility will have to go beyond earth to encompass the solar system; a fitting task in view of our 'cosmic connection', the scientific acceptance that even humans are made of the same stuff as stars. Space science emphasizes our innate relatedness to each other and to all things around us and therefore concerns 'our goals, our objectives and our relationships. I think this new capability should not provide us only with tools, it should also help to change the agenda of Man' (Pal, in Karnik (ed.), 1982, p. 11).

These new images reflect the factual unity of mankind and of our larger environment. They have also provided an opening we badly needed. For years now we have had to grapple with the concept of limits: Einstein set limits to the speed of light; Heisenberg set limits to human capacity to comprehend reality; the second law of thermodynamics seems to set limits to the evolution of the universe and of time. More recently, we have had to cope with the idea of limits closer to home: limits to growth and limits to the carrying capacity of earth. These limits are, however, off-set by the opening-up of outer space activities: an opening which is intellectual, moral and practical. Cosmology, ecology and psychology merge in the contemplation of our rocky spaceship and its inhabitants.

The opening-up is very obvious in one perspective: the explosion of knowledge. Writing in 1963 about the prospects for 1988, the political scientist Karl Deutsch said:

Already man's entry into outer space will have made the world of 1988 unlike all its predecessors. . . . Outer space will have begun to transform human civilization on this planet, not because it will have become a major field of colonization or control, but because it will have become almost inevitably a new and major source of knowledge. [Deutsch, 1963, p. 142.]

This is corroborated by a statement in a recent United Nations document:

The increase in man's knowledge of the nature, origin and evolution of the universe, of galaxies, stars and the solar system, has been so remarkable in recent years that today's ideas and understanding scarcely resemble those of a generation ago. Much of this increase in knowledge has been due to observations that were impossible before the advent of the space age. [UN Document A/Conf.101/BP/1, 1982, p. 62.]

Thus, the new space telescope to be launched in 1986 is destined to see 'the edge of the universe and the beginning of time'.

So, in a short time we have learned more than ever before about the gravitational and magnetic fields of the earth, about solar winds and particle radiation, about the moon and the planets of our solar system, about supernovae, neutron stars and black holes. Above all, we now have new images of the universe, and of our place in space and time. Instead of being subdued by a Newtonian picture of a universe sedately and mechanically moving as a clockwork, we now see a changing universe in full evolution, to the point that we seem to 'live in a relatively peaceful suburb of a quiet galaxy of stars while all around us, far away in space, events of an unimaginable violence occur' (Calder, 1975, p. 7).

Politics and economics of outer space

In cosmic terms, our galactic suburb may be peaceful but in Terran terms space activities, similar to atomic weapons, have provided the bases for new geo-political and military power positions. Man, who previously lived and acted in two dimensions, on or close to the surface of the earth, can now act in three dimensions. As early as 1955, two American political scientists analysed this new situation:

As man reaches upward to the outer atmosphere, new political problems arise, the nature of which we are as yet unable to grasp. Heretofore, the relations between nations and military forces were determined by the geometry of a spheroid's curved surface . . . Henceforth, international relations will be geared to the more difficult geometry of the interior of a large spheroid enveloping at its core a smaller and impenetrable spheroid, the earth. But even more confusing, the radius of the outer spheroid—symbolizing the aeropause of the altitude which man has reached at any given time—is expanding. The technologically most advanced nations will operate within the highest aeropause, while the spheroids circumscribing the aerial capabilities of the more backward nations will have shorter radii. Hence, in the future, the geometry of power will be described by several enveloping spheroids of different sizes . . . Truly, a new *Weltbild* is emerging. [Possony *et al.*, 1955, p. 125.]

Events have proved this analysis to be correct. Three hundred years ago, those who controlled the sea also controlled the world. Now, the key is control of outer space. It has been said that the quarter century of activities in space has had its most significant impact on the balance of power. Through the extension of military activities into outer space, the 'high frontier' has also come to signify the increasing risk of nuclear war, starting not on earth, but in space. Thus increasingly it is said that effective control of space by one state would lead to planet-wide hegemony: one country able to control space and prevent the passage of other countries' vehicles could effectively rule the planet (e.g. Deudney, 1983).

The new geometry of international relations can also be

seen from another angle. It has been estimated that well over 90 per cent of all space activities have been carried out by the two superpowers. Even though a number of countries, individually or collectively, are developing rocket launching capability, in the foreseeable future only a few countries will possess a major capacity in this field. Other countries that wish to engage in outer space activities of their own will therefore be dependent on the few that have developed rocket technology and thereby the launching capability required; at present they can choose between either of the two superpowers or the European Space Agency to which in the near future may be added China, India and Japan.

This shows one of the characteristic contradictions or ambiguities of the space age. The totally new factor, that man and his instruments can leave earth, means that, for the first time, earth is experienced as a planet, part of a wider new environment. To a Martian, we are all Terrans. But this new activity at the planetary level takes place within mental patterns and social frameworks that are based on older and now outdated principles. At the national, and even more at the international level, the entire social system is lagging behind these developments. We have not yet succeeded in creating truly planetary institutions. While space activities have caused the creation of new social forms, these have often served to increase the gap between the space powers and all other countries.

There is also another side to the knowledge we have gained through outer space activities: a space programme requires the contributions from and co-operation among a large number of scientific disciplines. According to one Russian view, it would be impossible to construct and launch space vehicles without better metals, precision instruments, electronic components and computers, alongside advanced research in chemistry, physics, astronomy, mathematics and many other scientific disciplines. In the opinion of American experts, the following wide scientific areas play a decisive role in space technology: biology, behavioural sciences, chemistry, geology, engineering, mathematics, medicine and physics. And the emerging materials science in space has required the joint effort of crystal growers, fluid physicists, biologists, metallurgists, physicians and thermodynamicists to exploit the micro-gravitational environment. Thus, the use of outer space, in the same way as the exploitation of the sea beds and the management of the environment, requires new kinds of interdisciplinarity.

This co-operation and integration in the scientific–technical area is matched by similar developments in industry and organization. The quantity and complexity of problems, the contributions required from numerous scientific and technical fields, the need for co-ordination between the institutions, governments, space authorities, industries, universities, laboratories and ministries—all these factors have resulted in new forms of social organization. Also, in a new methodology, a new manner of approaching problem complexes comprising a great number of previously unknown part-problems, of analysing, processing and solving them, arises. It is, above all, through space technology that new methods for the development and operation of large systems and new concepts such as systems analysis have become known. For the moon landing, Apollo 11 and its launcher represented eight million individual parts working together, and possibly the largest single technological enterprise involving the efforts of over half a million people.

The most important spin-off from space operations is 'systems engineering', 'the professional drills that firmly subordinate means to ends and thereby prepare the way to a more humane and less self-willed technology' (Calder, 1969, p. 13.) The optimism of this statement seems unrealistic with the growing concern over military activities in outer space. However, this new methodology has given the space powers a real advantage, that shows in the potential capacity of official institutions and of industry to handle complex problems set in other areas too. It has repeatedly been stated that the real gap between American and European capacity is not technically, but socially conditioned. It is not the famous technology gap that is decisive, but a project gap: there are in Europe no, or very few, common projects large enough and complex enough to force the use of this new methodology.

Beside the political, military and social significance, it is also possible to trace the importance of space activities through their economic impact. According to various estimates, some one and a half million people were active in the USSR space effort by the end of the 1960s, with about 1 per cent of the Gross National Product (GNP) at their disposal. At the same time, the American aerospace industry employed a similar number of people who earned some $49.2 billion in salaries. According to some calculations of the present level of expenditure, the Soviet Union devotes about 2 per cent of its GNP to space and the United States about 0.5 per cent of its GNP.

We already know that the application of space technology has important implications for terrestrial activities: for communications and navigation, for meteorology and cartography, for the monitoring of the environment. Beyond these direct benefits (often called 'pay-offs' in the technical jargon) there is also much discussion about the indirect benefits, the technical and economic value of space technology or rather of technology originally developed for space purposes, when applied in other areas. This spin-off effect often refers to the by-products of space technology.

Thus, the American space agency, NASA has, by law, not only the task of developing and using new methods and inventions but also of transferring them to private industry. Technical innovations developed in the context of the United States space programme are published and made available to industry through a special project. It has been stated that during the short period of 1958–65 some ten thousand new products were developed from the basis of space technology.

There are many examples: a biological telemetry system for astronauts is now used in hospitals for heart control; in medicine, new surgical methods through deep-freeze and micro-instruments have been introduced; in industry, new materials are used such as cold and heat resistant metals, and new metal mixes. In connection with space activities, entire new areas have opened up such as bionics—the use of electronic circuits in living organisms—or the manufacture of so-called cyborgs, cybernetic organisms. Other areas where space-derived technologies have been introduced are plasma technology, automatic pilot and security equipment for airplanes, new radar systems, as well as magneto-hydrodynamic systems and other methods for generating energy. Russian scientists have emphasized the importance of space activities in electro-vacuum technology, atomic energy installations and in the chemical industry, particularly with regard to new plastic materials.

The American consumer industry has obviously participated in this bonanza: components for radio and television receivers, synthetic foodstuffs, space suits for firemen, even teflon in cooking pans—all are by-products of space technology. It should not surprise us that 'space industries' have grown like mushrooms out of the American soil. In some newspapers it became a fine art to recognize in 'Space Materials, Inc.', the brave new shape of a former ladies' underwear business.

These factors have also led the American aerospace industry to diversify, ramify and penetrate new areas such as medicine, electronics, communications, metal manufacturing, mechanics and oceanography, just to mention a few. Integrated systems technology, which evolved in the space industry, is now applied to ship building, city planning and labour market regulations. Not only have efforts been made to shorten the gap between development and application, but also the time span between discovery of new knowlege and its integration into scientific and technical education.

However, the space enthusiasts have not had it their way all the time. On the contrary, achievements in space have not prevented severe criticism on earth, including the witticism from the science-fiction writer Theodore Sturgeon that 'NASA's most astonishing achievement was to make mankind's greatest achievement look dull'.

It is difficult to judge attitudes within the USSR but 'clearly, Soviet leaders have consistently placed a high priority on space activities, linking this exploration to the most important accomplishments and destinies of socialism' (Deudney, 1982, p. 9).

It is easier to follow changes in public opinion and in outer space policy in the United States.

During the early and middle 1960s expressed criticism of the space programme was relatively low-key. The objections to high costs, particularly for the manned space programme, could be countered by reference to the expanding economy, the perceived need not to lag behind the Russians, the military implications; or reference to the fact that the space programme in any case consumed less than American ladies spent on cosmetics each year.

However, criticism has in later years become more virulent, and this for scientific, financial and social reasons. Expenditure on the Apollo programme was seen as too high in relation to what many considered to be limited scientific results. These views were related to the changes in social values during the late 1960s and early 1970s: Why spend so much on space when other more immediate needs were not satisfied, such as environmental protection, urban renewal, poverty? These new attitudes also led to the recognition of the need to evaluate technological programmes in a wider social perspective. In 1972 Congress, with reference to the decisions required in such technologically advanced areas as outer space, decided to establish an 'Office of Technology Assessment' with the task of providing assistance to the decision-makers in the social and political evaluation of technical programmes. One result of this new approach was the refusal by Congress to approve the construction of a supersonic airplane (the French and British parliaments were not as well advised).

These changes in attitude and the deteriorating economic situation led to a crisis in American space and aeronautical activities. NASA's staff fell from the 1967 high of 35,000 to a mere 14,000 during the late 1970s. The budget cuts of the Reagan administration have led to 'a sense that the shuttle Columbia will be rising into orbit . . . from the ruins of the nation's civilian space program' (Wilford, 1982). Recently, however, there has been another change of attitude and policy which will influence the future of activities in outer space.

Future trends

To provide a background to an attempt to project future trends in space flight, it seems helpful to use analyses of space activities up to the present. Some observers have distinguished four periods in outer space developments:

1. The first steps into outer space represented by the launching of Sputnik 1 in 1957, the first exploration of the moon (Luna 1 and 2 in 1959), the first living terrestrial beings in space (the dog Lajka and the first cosmonaut, Yuri Gagarin).

2. The race for the moon initiated by the famous speech by President Kennedy in May, 1961 which defined the 'conquest of the moon' as a national objective and as a response to the Russian victories in the first round of outer space activities. The USSR managed to land the first vehicle on the moon (Luna 9 in 1966) but in 1969 more than five hundred million television viewers the world over could follow one of the most spectacular events in human history: a man—the US astronaut Neil Armstrong—stepped down onto the moon, an event succinctly described as 'a perfect last day of the old world' (Clarke, 1970).

3. The race towards the planets: the Russians seemed to focus on the inner planets through the landing of Venera 7 in 1970 and the first images of Venus reaching earth in 1975. The Americans performed the first Mars-landing in 1976 (Viking) and later the Pioneer spacecrafts have travelled to the limits of the solar system and have transmitted, by the hundreds, photographs of Jupiter and Saturn. Finally, on 13 June 1983, Pioneer 10 became the first man-made object to leave the solar system when it crossed the orbit of Neptune, currently the most distant known planet.

4. Concurrently, the exploitation of near-by space: immediately following the first exploratory phase, the space powers put into orbit around the earth space vehicles that were increasingly heavier and more powerful, stayed longer in space and served more purposes: scientific, military, navigational, informational and observational. In 1977 some 10,000 space objects had been logged; in 1983 there were 1,228 functioning space vehicles and some 3,400 pieces of man-made space debris in orbit around the earth. And in the 1980s, with the Russian long-term journeys in outer space

and the first flights of the American re-usable space shuttle begins the era of permanent occupation of the earth's orbit.

What about the future? At this moment, two trends seem crucial: first, the objective of a base for activity *in* space, for which the required instruments are an economic re-usable space transport system and the construction of space platforms and space stations; second, the increasing militarization of outer space.

According to a recent analysis, 'the manned space efforts of the United States and the Soviet Union are currently proceeding along parallel paths but are temporarily pursuing different short-range goals, goals which are based on the different needs of the two countries' (Oberg, 1981, p. 14). In this view, the United States, Canada and the European Space Agency are together developing a space transportation system, based on the space shuttle, and a family of rocket engines to be carried by and launched from the shuttle and the Spacelab research module. The immediate goal is to establish easy and economical access from earth into space and back, and to conduct frequent scientific sorties into orbit, with a large variety of scientific instruments. Such capabilities could, towards the end of this decade, lead to the creation of one or more permanently occupied space stations of considerable size.

In contrast, the Soviet Union is now developing a small basic space-station module first, in order to gain long-term experience in space operations with a limited amount of research equipment. Later in the decade, it is expected that the Russians will introduce re-usable systems similar to the NASA space shuttle but at present they are relying on tried and expendable boosters while concentrating their efforts on extending the capabilities of the space-station system itself.

This policy seems to correspond to a much-quoted statement by Brezhnev: 'Soviet science considers the creation of orbital stations with changing crews as the highway of man into space'. Already in 1974, the late Dr Boris Petrov, Chairman of the Intercosmos Council, USSR Academy of Sciences, listed three methods for the development of large orbiting stations:

(i) Space stations can be injected into orbit in a fully assembled state, as it is done today. Naturally, the performance of the carrier rocket puts limits on the station's size and weight.
(ii) Stations can be designed in modules, each of which is sent into orbit by a separate carrier rocket. The modules are then docked together to form the space station.
(iii) Smaller units, assemblies, equipment and instrument modules are put into an 'assembly' orbit by separate carrier rockets and assembly is done by cosmonauts and a special craft equipped for this purpose. [Petrov, 1974, p. 402.]

The Soviet Union seems to have worked steadily towards the establishment of a permanent, earth-orbiting space station. The Russians have maintained a more or less continuous human presence in space: thus, from 1978 to 1981 Salyut 6 was almost continuously occupied by at least a two-person crew, and in 1982 the cosmonauts on Salyut 7 set a new record for an unbroken stay in space and received two visiting crews from earth, each staying for about a week. All available information indicates that the Soviet space programme will push on with its goal of setting up a permanently-manned space station and for this purpose also develop a multi-use orbital spaceship like the American space shuttle. When that goal is achieved, there will 'always be people in space, living, working, observing . . . and speaking Russian' (Oberg, 1983, p. 600).

At this stage, it would seem that the USSR space programme is in better shape, at least in terms of objectives, than the American programme. True, the re-usable shuttles Columbia and Challenger have been a success—but only at the expense of cancelling other space projects such as planetary exploration and aeronautical research. This has gone sufficiently far for an editorial in the New York Times to pointedly recall that the 'S' in NASA stands for space and not just for shuttle (New York Times, 4 November 1981). In addition, the space community has been divided about the value of the shuttle. According to some, the shuttle is a technical success but a financial monstrosity, a costly mistake that gobbles up scarce billions yet offers few significant advantages in space beyond what can be accomplished with expendable rockets. One way for NASA to counteract these criticisms and the budget cuts is to interest other countries in co-operative ventures even though, finally, some competition is now offered through the relative success of the European launcher Ariane. Even the recent space policies announced by the Reagan administration are said to be the outcome of fierce struggle among up to nine involved agencies with conflicting interests. They are also said to represent important shifts in emphasis towards the increasing involvement of the private sector in space ventures and at the same time increased military control of activities in space reflected, for example, through the blending in the space transport programme of the previously separate civilian and military aspects. In this respect, observers point out, the United States and the USSR would now follow the same lines.

The programmes of the two space powers would also become more similar through the new American space policy formulated in 1982. It is announced that the new national space programme will include plans for a manned space station as one of the few new initiatives in the budget for 1985. Though the initial commitment would be at the relatively 'modest' level of $100–200 million, it might well represent the opening wedge in a programme that would cost $8–9 billion by 1991 and $20–30 billion by the year 2000. The manned space station is, however, no less controversial than the shuttle: NASA's space station concept has been described as more a means than an end, 'an orbiting white elephant unless its purpose were carefully defined' (*International Herald Tribune*, 29 December 1983).

There is no lack of plans and visions for the future. They include solar power satellites which will supposedly solve our energy problems on earth; space platforms and space stations of the most varied design and function; space operations centres which would include solar panels, habitat and workshop modules, fuel tanks and spacecraft hangars; unconfirmed reports of a Russian one hundred-man space station and, ultimately, space colonies in the shape of wheel-like toruses for tens of thousands of people. In this view, outer space is represented as the new 'high frontier' of human endeavour; these plans are seen not as a hazy vision but as firm proposals for the next steps into space which are being thrown open for public debate.

Low-cost space transport is seen as the key to a rich new continent in earth orbit: there are proposals for mining the moon and moving asteroids closer to earth for ease of operation; there is a growing interest in the advantages of undertaking industrial processes in the micro-gravity of outer space to manufacture drugs and specialized alloys, to produce semiconductors and grow silicon crystals. This push by private enterprise into outer space will, in political and legal terms, raise a host of fundamental questions of responsibility and regulation. There is in fact already controversy about the relative roles of public policy and private enterprise in respect of the future exploitation of the moon, similar to the controversies over the mining of the sea-beds in the negotiations on the Law of the Sea.

Some people even have grandiose ideas about space factories to process materials more efficiently than on earth, to manufacture totally new ones and to generate new industries not possible on earth. Most extreme is the idea of a third industrial revolution as the logical outcome of the first industrial revolution of the eighteenth century and the second, electronic revolution of the twentieth century.

These are the stepping stones that lead to the Third Industrial Revolution: the relocation of most industrial processes from the Planet Earth into the space environment where they will utilize the raw materials of the Solar System, tap the energy of Sol, and re-cycle the waste materials and energy back into the continuum of the universe. [Stine, 1974, p. 328.]

As could be expected, none of this is uncontroversial. These extreme ideas have been subject to sometimes fierce criticism. Some observers feel that decisions will have to be made on the simple question: will space become a source of, or, rather a sink, of wealth? Reality, though, is more complex and difficult than such formulas indicate. And before any of these plans can materialize we will have to cope with what the Secretary-General of the United Nations has likened to 'approaching storm clouds that threaten to cut off all rays of hope, in space and on earth': the increasing and rapidly escalating militarization of outer space.

The technologies of rocket propulsion and ballistic missiles have always been closely linked. The military, and particularly that of the two space powers, have since the beginning of the space age been heavily involved in outer space: according to some estimates up to 75 per cent of all satellites launched by the two superpowers have been devoted to military purposes. This use of outer space for military purposes has taken many forms: satellites for photographic and electronic surveillance, ocean surveillance, early warning and nuclear explosion detection, in addition to the uses of satellites for military world-wide communications, navigation, meteorology and geodesy.

There has, though, been a set of strange ambiguities about the military uses of outer space. The first ambiguity concerns the fact that space technology, in the same way as terrestrial technology, can be used for both civilian and military purposes: data on weather patterns, geodesy or world-wide communications can be made to serve both peaceful and military purposes. Secondly, reconnaissance and surveillance satellites have contributed to the reduction of tension and thus to stability, not only by guarding against surprise, but in the following ways: the agreement concluded between the two space powers to limit the number of strategic nuclear weapons would probably never have been concluded without the availability of such satellites that could be used to verify the compliance with the treaty, instead of controversial on-site inspection. The SALT 1 Treaty explicitly prohibits attempts to interfere with 'national technical means of verification' which include surveillance satellites as a main instrument. Thirdly, even though satellites have been used for military 'support' activities—mainly for what are called the 'three Cs': control, command, communications—for what has been defined as military but not aggressive purposes, they have still transformed space into a major frontier of strategic capability.

Thus, in the opinion of an informed analyst:

Satellites are so much part of modern weapons systems that they have been added to the already long list of potential military targets in any future war between the two powers. They are such important targets that the crippling or destruction of one or more satellites by the enemy would be tantamount to a declaration of war; war on earth would then start in space. Both sides have some conventional anti-satellite (ASAT) weapons and both are investigating other methods of destroying each other's satellites, including weapons which up to now were strictly in the province of science fiction writers. Among these are high-energy laser and particle beam weapons. [Jasani, 1982, p. 2.]

Thus, in recent years, the situation has changed for the worse. The new element which has caused increasing anxiety includes the development and, in certain cases, the testing of new space technologies whose functions are directly hostile and aggressive: anti-satellite weapons in the form of interceptor/destructor, or more popularly hunter/killer, satellites and other systems such as high-energy laser devices. The immediate effect of these developments seems to be destabilization, as reflected in a TASS news item stating that 'any impediment to the operation of surveillance satellites, and the more so, attempts to destroy them, might result in a situation where each side, deprived of reliable data on the opposing bloc's military preparations . . . could proceed from the "worst variants" ' (in *The Japan Times*, 8 July 1983, p. 4). Moreover, the development by the United States and the USSR, of satellites and systems capable of destroying satellites in orbit risks opening up space as another area for potential war. Of significance here is the statement by a high-ranking American military official that 'Space is not a mission, it is a place. It is a theater of operations. It is now time to treat it as a theater of operations' (quoted in Halloran, 1982).

In addition, there has been a profound change in American defence policy. In what has come to be known as the 'Star Wars' speech, in March 1983, President Reagan suggested that America should abandon the three-decade-old doctrine of deterring nuclear war by threat of retaliation and instead pursue a defensive strategy based on space-age weaponry designed to intercept and destroy incoming enemy missiles. The appropriately named MAD—Mutual Assured Destruction —would thus be replaced by what many observers see as an even more uncertain and threatening development. To many analysts, Reagan's dream of a 'truly lasting stability' represents a nightmare of a new and highly destabilizing space arms race in which the security of both nations will erode and their economies suffer, since the costs are seen to be in the $200–$300 billion range. Space wars can therefore not be seen as an alternative to war on earth, but rather as a prelude to war on earth.

The concern over the growing militarization of outer space was clearly expressed at the Second United Nations Outer Space Conference held in Vienna, in August 1982 (Unispace '82). The UN Secretary General, Javier Pérez de Cuellar, set the tone in his opening address when he spoke out against the increase in military components of space programmes as a frightening escalation of the arms race. He was followed by almost all delegations which finally managed to persuade the United States and some of its NATO allies to agree on a resolution recommending that this question should urgently be taken up by the UN Committee on Disarmament. In addition to the pressure for action within the UN context, there have been other reactions. The US Senate is discussing a resolution calling for immediate negotiations for a ban on weapons of any kind in outer space. The USSR has proposed a treaty for this purpose and Soviet President Andropov has also proposed a moratorium on the launching of anti-satellite weapons. Various national and international bodies are moving to promote a strengthening of existing international treaties so as to safeguard outer space for peaceful uses.

The challenge of space

We now face the real challenge of space—the critical choices confronting us. We can choose to act so that outer space becomes an added dimension to the growth potential of the human race, or we can choose to transform outer space into a new arena for our conflicts on earth. The easy way is to transport to outer space our fears and distrust. We could, on the other hand, take seriously not only the letter but also the spirit of the Outer Space Treaty of 1967 which introduces, for the first time, the concept of 'mankind' into international law, which internationalizes outer space and, despite its admittedly somewhat ambiguous language, is intended to establish a peaceful regime in outer space. Lately, however, the Treaty seems to be under such pressure that it is cracking at the seams: hence, the anxiety over killer satellites, space platforms created for military purposes and outer space laser weapons. Taking this road we would project into space both human self-destruction and human folly on a cosmic scale.

The other road is more difficult, challenging and requires an immense intellectual and organizational effort. Some of the benefits of outer space have already become obvious: the increase in knowledge, the advances in the monitoring of the environment, in navigation and communications, the possibility of new materials fabricated in outer space, even new energy sources. But here again, we face a choice: who will reap the benefits? Will it be, as with the sea-bed, only those countries that are already rich and technologically advanced? Will our use of outer space, even for purely peaceful purposes, do no more than aggravate terrestrial disparities? Thus, even to manage the resources offered by outer space, we need to put our own terrestrial house in order. So far, neither our thinking nor our institutions have been able to adapt to the new situation. Space activities imply a planetary perspective but they are still carried out within the framework of competing nation-states, as part of national jockeying for power and dominance, with glaring inequalities in capacity and advantage.

2 The communications revolution

A glance at history

Outer space activity would be impossible without the advances of rocket technology; it would be meaningless without the advances in communications technology. Communications satellites are developed at the meeting point of space technology and communications technology. It is the combination of both that has made satellite communications a key factor in the current communications revolution.

In historical perspective significant elements of this revolution appear quite clearly. For a larger part of human history, communications and transport were almost synonymous: their speed was equal as long as information had to be transported in physical form, from one place to another. Until the middle of the last century the communication of information over a distance was, with few exceptions, subject to the same constraints in time and space as had existed throughout history; not until the discovery of electricity did something new happen. One of the first practical applications of electricity was the telegraph, which represents the first step into the new world of telecommunications.

Once the first breakthrough had been achieved, new modes of communication followed at an ever-increasing pace. Following Bell's invention of the telephone in 1876, Marconi introduced, some twenty years later, a new and remarkable method of human communication: the use of electro-magnetic radiation without the need for physical support. The use of radio waves for communications provided the basis for the development of radio and television, for communication with ships and aircraft, and for satellite communication.

The recent counterparts to the early telecommunications services are numerous. In the late twentieth century they seem to produce fresh offspring every year, from teletext to telemetry, from facsimile transmission to videophone and home computer terminals. The telecommunications services are often described as the nervous system of modern society. Industry, administration, public institutions, social services, defence, education are all, to varying degrees, dependent on one or more of the mechanical and electronic descendants of the telegraph and the telephone. Many of these services are carried by electro-magnetic waves through wires and cables with ever greater capacity. More are carried through the air: 'air' in this case being the radio frequency spectrum, nature's vast but finite resource which cannot be subject to physical appropriation but requires agreement among users for its efficient use. It has rightly been said that the radio frequency spectrum has a similar relationship to telecommunications as the land has to farming or the sea to fishing. The increasing demands for more communication services and the introduction of new technologies result in growing pressure on earth's resources, including the radio waves propagated close to earth. Outer space thus appeared as a new and unexploited resource for communications.

Different communications services require different frequency volume, or bandwidth, depending on the kind, rate and volume of information to be transmitted. Telecommunications first used the lower frequency ranges which are easy and inexpensive in use but which only have limited information-carrying capacity. The history of telecommunications can be seen as a continuous effort to provide more capacity for information transfer, either through the development of new methods for a better and more economical use of the various frequency bands or by opening up higher frequency ranges which can accommodate larger volumes of information but also require a more sophisticated and therefore more expensive technology.

As the lower frequency bands became congested by traditional telecommunications services (mainly telegraphy, telephony and sound broadcasting), and as new services were introduced demanding more bandwidth (television and other video services, high-speed data transmission), the pressure for the use of higher frequencies increased. Satellites became a chosen instrument, not only because they can so easily span large distances but also because of the relative ease with which they can be made to operate in high and ever-higher frequency ranges.

Technical advances in telecommunications

As with any other new method for electronic information transfer over a distance, satellite communications form part of 'telecommunications', which is defined in a wide sense as 'any transmission, emission or reception of signs, signals, writing, images and sounds or intelligence of any nature by wire, radio, optical or other electromagnetic systems' (Final Acts, World Administrative Radio Conference, International Telecommunications Union, 1979). The development of communications satellite systems should therefore be placed in the context of technical advances in telecommunications generally. These advances are being made in three distinct fields:

— in the development of transmission systems that can simultaneously carry volumes of information which, by comparison with traditional methods, are enormous;
— in the construction of switching systems that enable the increasing number of messages to reach their destinations reliably and at an acceptable cost;
— in the manufacture of new devices operating in the home, office, factory, school and hospital that exploit the new and expanded services.

Satellites represent one of the technologies available for the transmission of information at a distance. The techniques now used include:

(i) Twisted pair wires, the copper wires which are traditionally used in telephone networks everywhere: this technique was designed primarily to transmit speech but can, with the addition of special equipment, also be used for certain other services that do not demand large bandwidth.
(ii) Coaxial cables which have considerably greater information-carrying capacity; coaxial cables are used in the telephone network and also in cable television systems for the delivery of colour television signals and other services to subscribers.

(iii) Microwave radio relay systems in the upper register of the radio frequency spectrum provide a major means for transmitting long-distance telephone and other telecommunications services, including the transmission of television programmes.

(iv) Optical fibres are thin strands of glass which can be made to carry lightbeams produced by a laser. Since lightwaves are higher in the frequency spectrum than radio waves, they have greater information-carrying capacity, to the point that their potential appears almost unlimited.

In contrast to satellite systems, all these transmission techniques require a path on earth, or near earth, whether messages are routed through wire or shunted along a series of microwave repeater stations. Thus, to extend a network, physical additions to the terrestrial pathways are necessary, be it underwater, or over mountains. To extend a satellite network, nothing more is required than to establish, at the desired location, an earth station which will be immediately capable of communicating with all other earth stations within the coverage zone of the satellite.

Another important aspect of telecommunications technology concerns the techniques which are used to process the information transmitted over the networks. In this context, the distinction between analog and digital processing techniques is of significance. The basic telecommunications infrastructure in place today is an analog transmission system originally designed to reproduce in a continuous manner the wave form of the human voice. Computers, however, function on the principle of representing information in discrete numerical form, i.e. as binary digital data. Following the introduction of computers and the perceived need to transmit digital data over telecommunications lines, digital transmission systems were found to have advantages in speed and volume not offered by voice-grade circuits. With the appropriate equipment, practically any analog signal can be converted into binary code, i.e. a set of binary digits or bits, and transmitted in digital form for reconstruction to the original form at the receiving end. The so-called bit rate represents another way of indicating the capacity required for the transmission of a particular type of signal: for a telephone conversation, 64,000 bits/second; for high-fidelity music, 320,000 bits/second; and for colour television, 92 million bits/second.

Digital transmission facilities are being introduced at an increasing rate in many countries and this trend is expected to continue so that, many experts believe, all major telecommunications networks, including satellite systems, will ultimately be based on digital technology. This would, however, require an upgrading of most conventional transmission facilities.

It would be of no use to increase transmission capacity without at the same time increasing switching capacity—although this mistake is often made. A switched communications network is one in which any user may be connected to any other user by means of a switch or sets of switches which establish the required routing path. The telephone network is the archetype of a switched network since the days of the old plug switchboard which, after various transformations, increasingly appears in the form of electronic exchanges based on computerized, digital switching. Without the computerized switching devices, whether located on earth or in a satellite, modern telecommunications traffic would simply grind to a halt. Modern switching techniques are a prerequisite for remote access to computers, and in most cases, for establishing two-way and interactive telecommunications links.

Many analysts envisage the telecommunications networks becoming fully digital, multifunctional information delivery systems which will be used for carrying all types of information—data, voice, text, picture, video—as encoded digital data, i.e. as a uniform stream of information. Consequently, such networks could provide a wide range of additional services as an integrated general computing/information/communication facility.

Thus, a new kind of telecommunications network is being promoted, an advanced information transport network that is designated as 'integrated services digital network', or ISDN. The objective is no less than 'the complete interconnection and interoperability of nearly all computer and telecommunication systems to provide universal and complete services for capturing, storing, processing, and transporting most of the information which society desires to retain or communicate' (Rutkowski, 1983, p. 14). The microchip and ISDN thus share the ambiguous distinction of being expected to transform dramatically all aspects of society.

All these new techniques and services depend on developments in electronics. Following the invention of the X-ray tube, the electron valve was, from the beginning of this century, the mainstay of electronics. Then, in 1948, came the invention of the transistor. Despite its small size, the transistor could do all that the electron tube had done—and do it better.

With the transistor, miniaturization began. It came just at the time it was needed, for computers and ballistic robots, for space vehicles and satellites. Miniaturization was a pre-condition for the explosion of the information industry.

About half a century lies between X-ray technology and the transistor—which itself, however, became outdated in a few years. Just as the transistor is superior to the electron valve, so is the integrated circuit superior to the transistor. And already miniaturization has itself been overtaken, by large-scale and very large-scale, or microminiaturization, by a variety of methods. The microelectronic push is transforming electronics and is rapidly penetrating computer and information technology, industrial processes and other economic activities, communications and entertainment, space and military operations. Microprocessors and microcomputers, based on the increasingly ubiquitous microchip, seem to spread like wildfire before we have even started to grasp their implications. They are heralded as the harbingers of decentralization, of a more equal distribution of 'information power' but the danger has already been perceived that the pursuit, to the exclusion of all else, of the rationalization processes made possible by microelectronics, will result in even more centralized and controlled patterns, de-skilling of jobs and deterioration of the quality of working life.

Characteristics of the communications revolution

It is difficult to find historical parallels for the current expansion in the range of techniques and devices in communications. Some compare it to the discovery of methods of cheap paper manufacture in the last century when each technical breakthrough brought about a new series of uses for paper—packaging, cheap educational materials, the mass circulation newspaper—and therefore fresh demands for raw materials and the search for new techniques of manufacture.

While there is a growing awareness that the convergent developments in telecommunications, computer science and audio-visual media will have a profound impact on society, there is less agreement on the significance of these developments. It would not be useful to repeat, once again, the list of technical devices that are being introduced or are under development in the laboratory. It is more useful to focus on some significant trends in technological development and their implications.

It seems to be an accepted fact that one of the characteristics of the new communications technology is a vast increase in the capacity to generate, store, process, transmit and receive information. The development of transmission capacity can serve as an example. In the 1930s, when modern cable technology was introduced through the coaxial cable, it was possible to provide for twenty-four telephone channels; these together, however, would not suffice for even one television channel. Current cable technology can provide 8,000 telephone channels, or eighty video-phones, or twelve television channels. There are plans for systems that can provide up to 200 television channels and work is in progress to use the visible part of the spectrum through optical fibres which can pass some 500,000 telephone calls or their equivalent in other services. Similar developments have taken place in radio communications, including space communications, through better use of lower frequency ranges and the conquest of the higher frequencies. In the development of cable and radio communications systems, the increase in capacity achieved during the last decades can rightly be described as sensational. Even more important perhaps are a number of other trends which have not been given the same attention.

First, there is flexibility and opportunity of choice. This works in several ways: we now have available a number of methods to provide a given service where previously we had one—or none. For the distribution of television programmes we can now choose between traditional terrestrial broadcasting, cable systems, satellites or the mail for recorded programmes, and combine these methods in various patterns. Similarly, the transmission of voice, telex or data, can take place through twisted pair copper wires, coaxial cable, satellites, optical fibres, or ordinary mail for hardcopy or recorded output. Originally, 'news' was distributed orally; this was followed by writing and print, then by film, radio and television, with each medium presenting the news according to its own characteristics and constraints. These patterns are now changing: 'printed' and graphically-presented information is regularly distributed electronically and the printed physical form may be reconstituted at reception.

This diversity also functions in a different way. Traditionally, the television set has been used for one purpose only, the reception of over-the-air television broadcasting. At present, new uses are being added at great speed: reception of cable or satellite transmissions, playback of videogrammes, electronic games, new two-way services in combination with telephone or broadband networks, display of computerized data, surveillance and control of production processes or traffic patterns. The television set should more accurately be described as a multipurpose display unit, but its definition and legal status is essentially tied to the one original function. [Ploman and Hamilton, 1980, p. 151.]

Second, there is the possibility of combining different, and so far separate technologies, systems and services. A broadcasting network can now combine over-the-air transmission with cable and/or satellite transmission. Similarly, the telecommunications networks now use a variety of transmission methods for telephony, telex, data and other communications services. Even more striking is the emergence of new services which result from the combination of various technologies. Such new systems for the transmission of information as teletext and, even more so, the interactive videotex systems, combine features from telecommunications, broadcast and computer technology. This trend towards new combinations is related to a trend towards convergence, even integration of techniques and services. At the technical level, one example is the introduction of digital message processing and transmission techniques; at the level of communications systems, the corresponding development is towards integrated networks which are capable of carrying any kind of message for transmission and reception in any desired form. Thus, new forms of communications and information transfer are introduced in patterns that have not been foreseen in policy, structure or national legislation.

One aspect of these trends of major importance is that technology has now reached a stage which makes possible almost any desired configuration. If decisions are to be made on the basis of what is technically feasible, economically reasonable, and socially desirable, it is the technical part which causes the least problems. Consequently, economic and social considerations take on greater importance and have proved to be more difficult of solution when it is no longer possible to hide behind technical constraints.

One attempt to solve this issue has been by the approach of 'intermediate' or 'appropriate' technology. In general these approaches have come to be associated with the movement that has taken as its motto 'small is beautiful'. Admittedly, technical and economic involvement has been too highly focused on the large, heavy, high technologies but there is reason to recall the original meaning of the word 'appropriate': that technology—whether low or high—which best corresponds to all defined requirements and assessments, is the appropriate one. There has been controversy over the resources allocated to satellite communications, as an example of high and centralized technology, which is therefore inappropriate in the context of developing countries. Obviously, each technology carries its own bias, rooted in the economic and technological environment from which it originates, and thereby also expresses a specific organizational and structural approach. This makes it even more important to assess, as far as is possible, any new technology in terms of its implications, so as to decide, from an informed viewpoint, first the 'if' of its introduction, then the 'how' and the 'when'. Thus, the starting point should be: not a particular technology or system, but the needs and requirements in a given case, the options available, the organization of use, and the relative costs—economic as well as social—to society and to users.

A third trend might be seen under the sign of contradictory developments (in opposite and not necessarily complementary directions). The trend toward overall, integrated configurations meets another trend in the opposite direction towards decentralized and individualized concepts and patterns of use.

This trend is obvious in the demands for localized broadcasting and in the activities of video-groups who have taken the electronic media out of the mass media sphere and transformed them into an individual or group medium of expression, similar to what had already happened in the cinema. Individual choice is the main argument for the video-cassette and the videodisc: choice of timing and choice of content. Remote access to computerized data banks could serve the same purpose of allowing the individual user to control his search for information. [Ploman and Hamilton, 1980, p. 152.]

The possible scope now ranges from the most extensive, world-wide patterns (e.g. international satellite systems used to interconnect national and regional communications systems) to localized or community activities and individual use (cf. centralized computerized data systems and personal microcomputers). This corresponds to the contradictory fears expressed at the same time: fears of a corporate global village and Big Brother wrapped up in one, and fears of the social fragmentation implied in exclusively individual or small-group or minority use of technologies and media —everybody 'doing their own thing'.

Fourth, there are few better yardsticks of the inequalities that exist between different sections of the human race than the present differences in provision of telecommunications services. Take the telephone, for example, which was invented more than a century ago but which has spread very unevenly around the globe. In the whole world there are some 358 million telephones; but only 30 million of them are in Central and South America, Africa and Asia put together. In parts of South America and Africa there is less than one telephone to every two or three hundred people; in Sweden there are nearly sixty for every hundred. The other major form of public telecommunications, broadcasting, is expanding in a similar way to telephony. The growth is explosive yet uneven. There are some 25,500 fixed radio transmitters in the world. Only 6,600 are located in developing countries. In North America the radio-receiving population has surpassed the human population—1,790 sets per 1,000 inhabitants —while in developing countries there is an average of seventy-six sets per 1,000 inhabitants. The discrepancy is even more obvious in the case of television. There are almost 24,000 television transmitters in the industrialized countries, while the rest of the world has only slightly more than a thousand. On this basis, the communications-rich countries are able to acquire ever newer, and more, information, entertainment and education facilities while the communications-poor countries face the struggle to provide even basic services of communications.

All societies now find themselves at the brink of a vast and unprecedented expansion in the number of services available through telecommunications. Yet most countries find it difficult to sort out their policies, to weigh the relative demands between different services, to relate telecommunications to other systems used for the transport of ideas and of people, and to choose efficient social instruments for formulating policy and operating the different systems.

Broadcasting, uneasily perched between telecommunications, journalism and entertainment, provides an example of this confusion. Throughout the world, it was the characteristics of a particular technology, i.e. over-the-air transmission, which were used as the basis for its establishment, for the institutional structures and the legal framework. Thus, broadcasting was made to follow telecommunication patterns rather than those of either newspapers or the cinema— or a new pattern altogether. Now, with the coming of satellites, cable television and other new technologies, new systems, new ideas, the broadcasting institutions have become involved in new services, in new social, cultural and legal issues— no wonder that they feel themselves in a state of continuous and confusing change.

The fauna of communications services once had a commendable simplicity. The small number of species could be grouped easily into a few neat categories. Pride of place was given to the print media; next came the telephone and the telegraph, which were regarded as belonging to the separate world of telecommunications, as were the more specialized maritime and navigation services.

Cinema was different yet again—a public spectacle. Radio and television were subsumed under the new umbrella of broadcasting. The radio amateurs lived somewhere in their own corner. Telex, mobile radio communications and other newcomers did not attract the public interest they merited since they seemed tied to the corporate world.

We now face a strange bestiary: the most obvious feature of the present communications scene has been the proliferation of unconnected gadgets which clutter and occasionally decorate our environment—telephones, radio sets, record-players, telex printers, television receivers, microfiche readers, video-recorders, photocopiers, data terminals and pocket calculators. Generally, each piece of equipment has been used for one purpose only: each communication service was associated with a special transmission or distribution system and developed within a specific institutional framework with its own mode of financing, set of policies and legal framework. Each species was seen as inherently distinct in terms of the technology used, the services provided and the regulation required.

Current changes in technology have made possible new patterns, traditional services being combined with new ones into unexpected hybrid shapes and uses, in defiance of established categories. The marriage of the telephone and television begets the videophone; the letter is supplemented with telefacsimile or messages recorded on sound cassettes; the cinema goes to the home and the videotape to the cinema. Traditionally, radio was intended for mass audiences and the telephone for two correspondents: both are now being used for and by groups of varying sizes. Artists work at electronic consoles. The written or printed word loses its physical form and is transformed into an insubstantial flicker on a visual display unit. The combination of the telephone, television and computer systems provides for unprecedented services which break the barrier between print and electronics.

New devices are constantly being 'offered' to society by technology. Some of them are deliberately tailored to fill human and social needs, such as education or business communication; others are developed by the electronics industry in the hope that they will 'catch on' with the public and become desirable instruments, such as the videocassette.

Industries make guesses and judgements as to what may be required, and what pace of development. Societies too must make judgements about their evolving needs.

The sheer range of new services and their wide social applicability— and implications—make it even more urgent for all countries to discern clearly their own needs over the next decades. Societies engaged in planning their future communications systems are now able to consider new options in the use of available resources. [International Institute of Communications, 1977, pp. 2-3.]

The emerging information society

With the advent of electronic communications, many analysts postulate a new communications and information environment whose invisible and often transnational networks constitute a radically changed global context for economic and cultural life. The movement towards electronic information devices is seen as the fourth major revolution in communications, comparable in impact to the development of speech, the introduction of writing and the invention of printing. In a larger perspective, the current revolution in communications and information is comparable to the transition made thousands of years ago from hunting to agriculture or, more recently, from agriculture to industry. This perspective provides the much-discussed sequence of a move from an industrial economy based on the exploitation of energy, to a post-industrial economy focused on information and knowledge as main resources.

It has almost become a cliché to speak of the 'communications revolution' or the advent of what has variously been called 'the information economy' or the 'information society'. Whatever else these expressions might indicate, they point to a fundamental change in our attitude to the entire communications/information complex. Despite the variety of approaches, there seems to be agreement on at least one point: communication is a prerequisite for all social organization—whether among animals or humans. Without communication there can be no community. In this perspective, all societies have been information societies. What we tend to forget is that this view of society is recent. Clearly, throughout human history, communications have retained the attention of rulers. All societies have evolved practices for organizing their information flows, and all have accumulated stocks of knowledge. However, communications, information and the application of knowledge do not seem to have been perceived as problems *per se* but rather to have been taken for granted, seen almost as natural constants, placed in the service of other social activities and goals whether they were expressed in religious, political or economic terms. It is only in recent years that there has occurred a shift in perspective so that communications and information by themselves have become issues in society, at the level of individuals, groups and institutions.

Analysts have pointed to the degree to which issues of communications and information have invaded the social, economic and political agenda, nationally and internationally. Thus communications and information both reflect and originate processes and phenomena which we obscurely but deeply feel are changing society and the outlook on society, as still dimly understood re-orderings of economic, social and cultural patterns.

This fundamental change in attitudes towards the communications/information complex is so recent that we still seem unsure about what it means and how to cope with it. We face a paradoxical and contradictory situation. On the one hand, each scientific discipline, each profession has developed approaches and paradigms of its own, as has each government department. Social responsibility is ill-defined and dispersed. On the other hand, more coherent, even holistic approaches are emerging, which have resulted in a new way of looking at communications and information: communication patterns and information flows are now used as basic concepts in the interpretation of natural phenomena (e.g. genetics) and of social organization; they are seen to affect social relations within countries and behaviour among nations, the nature of work and leisure, patterns of spatial movement and arrangement, the processes of learning, knowledge and culture.

Recently, however, the often euphoric or deterministic approaches to these developments have been questioned. We have been warned against the myths of the electronic revolution, and reminded of the inherent limitations of communications for the solution of urgent problems and of the risk of replacing 'economic man' of ill-fame by a new reductionism in the form of 'communications man' or 'homo informaticus'. Disquieting questions have been raised about essential aspects of the emerging information society: what are the limits—biological and social—to human information handling? Has the information explosion tended to reinforce the fragmentation of knowledge, resulting in growing cognitive dissonance and in disjointed images of reality? Will present trends, without corrective measures, replace the haves and have-nots of industrial society with the knows and know-nots of information society?

This perspective does, however, represent only one dimension of the emerging information society. There is another trend, also representing a complex set of developments: this includes the resurgence and growth of local, regional and national self-assertion, expressing itself, for example, in an insistence on the recognition and use of local languages and even dialects. It manifests itself in the emergence of new issue-oriented groups and in grass-root movements becoming increasingly vocal and visible both in developing and in industrialized countries.

Crucial goals of these movements concern communications and culture in the form of demands for a greater measure of self-control over communication systems and information generation. These demands are reflected in the clamour for greater participation and access, for a more equitable distribution of communication facilities and information resources, within and between societies, in the introduction of new concepts such as the right to communicate, and in such phenomena as the underground press, 'guerrilla television', demonstrations, sit-ins, terrorism, street theatre and other ways—traditional and novel—of exercising what might be called the right to be heard by the 'unseen voices'.

The new electronic technologies are therefore seen as technologies of knowledge and of organization: they would have a profound impact on virtually all aspects of society since they affect the manner in which we organize our activities—whether it be through office automation, computer-assisted art, or a profound restructuring of industrial processes. The changes in modalities of communication and information handling are crucial in their impact on cultural forms and identities. Inherent in these changes is a shift from the traditional Western paradigm of linear communication to modes of communication that are multimodal, inter-active, potentially more participatory and also, perhaps, more congenial with non-Western forms of knowing and communicating. The fusion of technologies changes the relationships with the social and physical environment and, most importantly, reshapes the information content and the symbol systems that sustain all cultures.

3 Space communications and communications satellites

Outer space activities and communications are very closely linked: without communications, space activities would be meaningless and often impossible. Communications are therefore an integral part of outer space activities whether we are dealing with the exploration of the solar system or with information transfer closer to home.

In terms of purpose, space communications present two different aspects. On the one hand, space communication is part of space exploration and serves to establish communication with space vehicles, for telemetry, telecommand and for the transmission of data back to earth. On the other hand, space communication technology is used for purposes here on earth and this is the aspect on which this study will focus.

These earth-oriented space communications services can be classified in various ways. The American space agency NASA used to indicate four major groups of activity: communications and navigation; meteorology; remote sensing, and geodesy. From a technical-administrative perspective, the International Telecommunication Union defines a series of satellite services such as the fixed-satellite, mobile-satellite, broadcasting-satellite, earth-exploration, radio navigation and still other satellite services. At the recent UN Outer Space Conference, Unispace 82, the following current and potential applications of space technology were discussed: telecommunications, mobile communications, land-mobile communications, maritime communications, aeronautical communications, satellite broadcasting, remote sensing, meteorology, navigation and geodesy. The easiest way to define these would be to use two major categories:

— *Observation satellites*, that are used for the collection, and increasingly also for the processing of data and the transmission of information to earth.

— *Communication satellites*, that are used for the transmission, distribution, and dissemination of information from and to various locations on earth.

Observation satellite systems

Outer space gave us the first picture of our planet. Already, at the first UN Outer Space Conference in 1968, the use of satellites for monitoring the earth's environment was recognized. The importance of this new capability is its timing, which is tied in with the growing awareness of our ecological mismanagement of the earth. Following the UN Conference on the environment held in Stockholm in 1972, and the first Club of Rome report on 'Limits to Growth', the decade of the 1970s witnessed an increasing recognition that the resources of the earth are not limitless: if even the basic needs of people on earth are to be met on a continuing basis, the earth's natural resources will have to be managed with care and foresight.

In the past, when human impact on nature was relatively limited and when most production and consumption were on a local scale, an 'intuitive' understanding of the immediate environment seemed sufficient to sustain humanity. Now, with human power over the environment increasing and with an economy based on international exchange, the traditional knowledge that shaped human attitudes toward the planet is no longer adequate. If man is to acquire a new understanding of his planet he needs new tools. And if this understanding is to be dynamic and global, the tools must be of a global and dynamic nature. To many observers, the potential benefits of applying space technology to the monitoring of our environment are such that earth resources satellites are seen as 'rescuing earth from outer space' (Siuru and Holder, 1972).

The first views of earth from outer space, obtained about 20 years ago, of its active atmosphere, and of its variegated land and seascapes were exciting, provocative and difficult to use . . . The wealth of potential information was suspected, but it took man nearly a decade to appreciate the amount of information available and to learn how to use it. [UN Document A/Conf. 101/BP/3, 1982, p. 5.]

Why space? What is so special about surveying earth from space? Are not aerial photography and ground surveys sufficient? Satellite technology has certain unique characteristics. In principle, a satellite circling the earth at an altitude of several hundred kilometres and above or in a geo-stationary orbit at an altitude of 36,000 km can do certain things better than any other method of survey. First of all, a satellite can see a vast area at any one time. Thus, a satellite travelling at an altitude about seventy times higher than a commercial jet plane can cover in a single photograph a large portion of the landscape. Whole mountain ranges, such as the Himalayas or the Andes, can be seen at the same time under the same light conditions. In a single picture one can trace a water course from mountain brook to ocean.

Because a satellite is so high above the earth, every point on the earth below appears to be on a flat surface perpendicular to the view from the satellite. Thus, a photograph of the earth serves as a map with very little distortion. In this fact lies great hope for the map-maker and the many planners of resources who rely on maps. It has been estimated that even after fifty years of aerial photography only about 30 per cent of the earth's maps can be considered adequate. Also, satellite mapping is faster and cheaper than traditional methods: a satellite map of the United States would require some four hundred photographs, which could be compiled in a week. In contrast, mapping from an aircraft would require about one million photographs, which it would cost $60 million to compile.

Satellites can also survey areas of the earth which are virtually inaccessible by any other means. This is an especial advantage for oceanography and to gain an understanding of the way in which global weather circulation patterns are affected by changes in the sea surface temperature. Most of the expanse of the oceans are rarely seen by man and it is easy to imagine the staggering fleet of ships and aircraft that would be needed to monitor the oceans and in a manner that can easily be achieved from a satellite.

A satellite traces a path on the earth that is repeated every so many orbits. If the right orbit is selected the satellite will even pass over the same location on the earth's surface every couple of days or weeks at precisely the same time each day. This function is particularly useful if we want to study a natural phenomenon through changes in time—for example the changes in crops through the growing season, or the build-up of water into a flood.

Also, satellites can operate for long periods of time at little or no additional cost once they are in orbit. An aircraft has to be refuelled every time it lifts off; for a satellite the only additional cost is in handling and processing the data on the ground.

The pictures of earth taken with hand-held cameras by astronauts and cosmonauts from the early manned satellites were dramatic but had limited application. Soon multi-spectral cameras were designed to use a variety of films and filters to allow more flexibility in image processing. While satellite photography gives important information it will not tell the whole story. Other types of 'cameras' or sensors will provide much additional data that is not visible to the human eye or to the ordinary camera. Every piece of matter, whether it be a plant, an animal or a lake, reflects energy in a unique way: this is called its 'signature'. Not only is this signature unique to an object but it changes as the condition of the object changes. For example, healthy crops or forests have a different signature than diseased ones, as does salt, as opposed to pure water, and polluted air as opposed to clean air. The problem is to pick the right sensor for the pheno-mena or event we want to monitor and study. These sensors, with the ability to measure miniscule amounts of energy, range from cameras using infra-red film to radar and micro-wave transmitters and receivers.

Thus, when a forest is filmed by an infra-red camera, the healthy trees appear red but the sick ones show up blue. On infra-red film, hot, polluted water entering a cool river is clearly visible. By filtering out certain colours from images recorded on photographic film, crops of barley can be dis-tinguished from fields of oats and the depths of coastal waters can be measured.

Remote sensing technology with its various components can therefore be described as a technical imitation, modification and extension of the eye/brain system. The satellite can also function as a data collection system. Sensors placed on the ground transmit information to the satellite which in turn retransmits this information to ground-based data collection centres. Such small environmental sensing stations are called remote data collection platforms and are found on ships, remote weather stations, buoys, agricultural stations, fire warning stations, seismic and seismic sea wave stations.

The large amount of data that can be made available through remote sensing via satellites makes the handling of these data and their translation into usable form one of the most important and most difficult aspects of satellite operation. The photographic data produced can be interpreted using techniques of air-photo interpretation; alternatively, the data can be analysed more precisely and quantitatively using digital data processing. Computers can assist in the interpretation in various ways and since this interpretation technique is still largely experimental, new methods are being developed. The simplest and least expensive method is to use the computer to optimize or enhance the imagery for visual interpretation. The standard photographic products are often far from ideal in lightness, contrast, resolution and spectral balance. Special processing can be used to produce an image that is optimized for a specific application. Computers can also be used for analysing images on the basis of the image brightness in each of the spectral bands.

With the new equipment used in earth resources, meteorological satellites and in other space missions, it has become possible to trace the wanderings of one particular shark, suitably marked, to identify one thousand year-old Maya canals hidden in the jungle and even to see under the skin of the earth, as in the discovery of 40 million year-old dried-up river basins in the eastern Sahara. And the French satellite SPOT is expected to provide images of areas down to ten metres across and to provide accurate identification and mapping of all major streets, large buildings and parts of a city.

The technology and equipment that were successfully tested during the early stages of space exploration are now finding wide applications in the global observation of earth:

of the atmosphere, the oceans, the landmasses, and, for a quite different purpose, in military reconnaissance in all three realms.

Meteorology

International co-operation has been most pronounced in the field of meteorology, which by definition is an international science. Satellites play a key role in the international programmes set up by the World Meteorological Organization. The space-based sub-system of the Global Observing System (GOS) consists of five geo-stationary meteorological satellites launched by the United States, Japan, ESA and a system of polar-orbiting meteorological satellites operated by the USSR and the United States.

Data received from these satellites can be used to determine profiles of atmospheric temperatures, cloud cover and height data, snow-melt, surface water boundaries, water vapour sensing, with greater atmospheric depth and atmospheric ozone data, important for environmental monitoring. Such information constitutes a major aspect of the international programme initiated in 1978 under the title of the Global Atmospheric Research Programme (GARP). The goal of this programme is to increase understanding of atmospheric processes and help improve forecasting ability. These satellites are also currently being used as part of the World Weather Watch (WWW) in monitoring and tracking typhoons and hurricanes in the seas and oceans of the world and in keeping watch on the desertification effects of droughts, particularly in the Sahel region of Africa.

Meteorological satellites are often capable of 'seeing' much more than clouds and weather systems. City lights, fires, air and water pollution, aurora phenomena, dust and sand storms, snow cover, ice formations, ocean currents and energy waste form part of the environmental information provided.

Remote sensing

While remote sensing of the environment uses data from a variety of specialized meteorological and oceanographic satellites, a major role has already been played by special earth resources satellites. Since 1972, experimental satellites such as the American Landsat 1, 2 and 3, and the Russian Soyuz 5 and 6 and experimental space platforms such as Skylab have collected data over most parts of the globe. These data have found applications in agriculture, forestry and range management, hydrology, geology, oceanography, energy resources, environment, flood and disaster warning, land use, mapping and charting.

Applications of satellite remote sensing have been largely developed in the industrialized countries. Even though these applications need to be adapted in order to be of use in other regions, over eighty developing countries have used remote sensing data from Landsat. The value to developing countries has been recognized in particular by India, which in 1979 had its own satellite for earth observation, launched by the Soviet Union. This satellite is particularly useful for the study of those resources that change with time and are renewable, such as cultivated land, forests, rivers and wetlands in coastal areas. India is now working on an operational remote-sensing satellite system which will be applicable mainly to agriculture, soil studies, hydrology, forestry, coastal oceanography and geology.

Remote sensing via satellites has a difficult problem in that it has not yet found its proper institutional setting. According to an analysis of the American situation:

Landsat, unlike the weather or communications satellites, has no natural home. The market is inchoate and fragmented. Individual users may be enthusiastic, but they are scattered through hundreds of federal agencies, universities and private industries. There is no one entity that caters to all these interests. Worse, Landsat imagery serves both the public welfare (as in pollution monitoring and land use planning) and private profit (as in mining and oil exploration). People cannot even agree on whether Landsat should be operated by the public sector or the private sector. [Waldrop, 1982, p. 40.]

In keeping with its general economic policy, the Reagan administration seems intent on selling the American earth resources satellites to private enterprise, perhaps even the weather satellites. These proposals have caused considerable controversy in scientific, business and political circles. It seems that members of Congress have been besieged with calls from worried farm, aviation and other groups who feel that weather forecasting is, after all, a legitimate public function, nationally and internationally. At the prompting of Congress it was agreed that the American weather satellite systems will remain in the public sector, but there is still uncertainty about the remote-sensing satellite systems. It might therefore well be that other countries such as France might beat the United States in the race to establish an advanced earth resources satellite system with the objective of providing relevant data for sale; the French SPOT satellite, undertaken with certain involvement from other countries—mainly Belgium and Sweden—is scheduled for launch in 1984.

Thus, the international situation is unclear, to say the least. The international remote-sensing satellite scene during the 1980s will be dominated by the Landsat 4 and its successors, various USSR satellites, the SPOT series, radar satellites from Canada and ESA, a number of Japanese satellites, the Indian programme and those planned by China, Brazil and perhaps some other countries.

There is no doubt, therefore, that there would be more than adequate data gathering capability in the second half of this decade, and yet there is no guarantee provided by any operator that such systems would continue to be available to all countries at reasonable prices. All the systems mentioned so far are still considered experimental or pre-operational because the methodology of charging for the data and assuring continuity of service has not yet been established. . . . We are truly in a peculiar situation. [Pal, 1983, p. 7.]

Many observers cannot see why it should not be possible to establish an international earth resources system when we have managed to find ways to operate international communications and navigation satellite systems. In concrete terms, it has been proposed that remote-sensing systems, advanced satellite communications and computer networks using platforms in geo-stationary orbit could be very useful. Thus, 'the crisis of aspirations that denies the world's citizens the right to expand and grow through an improved understanding of "this Island Earth" can be aided through the merger and advancement of the synergistic technologies of telecommunications, computers and Earth observations.' (McElroy, 1983, p. 4.)

The techniques used for earth resources and meteorological satellites have found another and more controversial use in reconnaissance satellites. Since the 1960s the United States and the Soviet Union have kept watch on each other from space, using reconnaissance satellites to observe ballistic missile sites, weapons testing and deployment, military exercises, factory and shipyard construction and a whole range of other 'interesting' activities. This may seem like an aggressive and unwarranted intrusion into national security but, as mentioned earlier, it is to some extent a part of the balance of power. The importance of these reconnaissance satellites and their findings has been implicitly acknowledged by the inclusion in the SALT agreements of clauses specifically prohibiting interference with the other party's 'national technical means of verification'.

Likely further developments include the extended use of photo-interpretation techniques by a large number of users, particularly in developing countries. The use of computer processing, interactive and automatic analysis and the integration of information into computer banks is expected to grow rapidly. Data from a variety of remote-sensing systems, along with meteorological, climatological and hydrological data, and economic, social and demographic statistics may be combined in a common geographic framework.

Technology should serve human needs; human needs should not be sacrificed to the requirements of technology. The development of earth observation satellite systems and their application will therefore depend on the decisions that are made as to how the earth's resources are to be managed and what will be the extent of international co-operation. Traditionally, human activities were planned and co-ordinated on a local or national basis so that the information on the state of local and national resources is updated at intervals of five to ten years at the best. The effective use of satellite data both permits and demands new approaches to national and international resource management. If careful and planned management in a context of international co-operation is made a priority, then satellite systems for earth observation should develop rapidly and should make an important contribution to human welfare.

Communications satellite systems

It is common practice to trace the history of communications satellites to a specific point in time and to a specific person: the date is October 1945 and the person Arthur C. Clarke, then secretary of the British Interplanetary Society, a position which at that time enjoyed a modest, if not dubious reputation. Clarke published in a specialized magazine called *Wireless World* an article which was first given the title 'The future of world communications' but was later published under the title 'Extraterrestrial relay'.

The article begins with a discussion on the use of radio and television over long distances. Emphasis is put on the fact that television transmissions require the installation of expensive coaxial cables or radio relays but that none of these methods could be used for transatlantic transmissions. Clarke then proposes the use of artifical satellites which, at a sufficiently high orbit of 36,000 km, and with a given revolution, would take twenty-four hours to circle the earth once so that, positioned over the equator, they would appear to be stationary in relation to a given point on earth. Clarke envisaged a manned space station which would be provided with electronic equipment for the reception and retransmission of all types of radio signals from any point in the area it covered. Since one station would not be sufficient to cover the whole earth, Clarke proposed to launch three, equidistant satellites over the equator.

Since then both Clarke and communications satellites have taken many steps forward. Clarke has become known as an author of scientific and popular science works, and of science fiction, as a speaker at conferences, and has been given medals and awards. His fame increased as the co-author of the film *2001*. Many times, Clarke has rather wistfully complained that 'it is with somewhat mixed feelings that I can claim to have originated one of the most commercially viable ideas of the twentieth century, and to have sold it for just $40' (Clarke, 1967, p. 119). He was consoled to some extent by the knowledge that in 1945 such an idea could not have been patented since it was not then technically feasible.

Development of satellite technology

Communications satellites also developed in a very different way from that proposed by Clarke: the miniaturization of electronic equipment made it possible to use unmanned automated satellites. Apart from the rocket technology which made possible the launching of satellites, it was the development of electronics which provided the means for the manufacture of efficient satellites and ground equipment: transistors, solar cells, noise-reducing masers, and parametric amplifiers, travelling wave-tubes with long lifetimes, threshold-reducing FM detectors and other esoteric devices. Equally necessary was the use of advanced computers and electronic data processing in order to define the orbit, steer the satellite and direct the antennas of the early earth stations.

In the United States information technology was prepared for outer space by the time the first Sputnik was launched in 1957. In fact, space communications had already begun during the previous decade: the moon had been used as a reflector for messages from earth. Following the first radar contact with the moon in 1946, the US Navy used the moon as a communication station between Washington and Hawaii, from 1959 to 1963. These operations developed a new idea: the orbit of the moon is very awkward for the establishment of communication links between various points on earth —why then not use a large, controllable, artificial balloon for the purpose? And so the first use of a satellite in the transmission of voice and picture signals from one place on the earth to another was the Echo project, in which signals were bounced off a large aluminized balloon. Echo 1 was used to relay telephony, facsimile and data. The Echo satellite was visible to the naked eye and was followed with great— and continuous—interest since it functioned satisfactorily until 1968. It was followed by a larger version orbited in 1964.

These were passive satellites in that they simply acted as mirrors for radio signals. The first active satellite was Courier, launched by the United States in 1960—the third anniversary of the start of the space era. An active satellite is one whose communications package, or repeater, receives the signal from earth, translates its frequencies as required, amplifies and re-transmits the signal to earth. Apart from the Echo experiments this is the type of satellite now used exclusively in communications applications. Since Courier—a US Army project—there has been a veritable explosion in communications satellites.

It is of interest to highlight some of the 'firsts' in the development of communications satellites:

(i) Oscar 1, launched 10 July 1961, was the first satellite devoted entirely to amateur radio.

(ii) Telstar 1, launched 10 July 1962, provided the first transatlantic relay of television signals, the first relay of colour television, and performed tests of broadband microwave communications via space.

(iii) Syncom 2, launched 26 July 1963, provided the first successful relay of communications via a satellite in geo-stationary orbit. Up to this time satellites had been placed in 'near-earth' orbits and were therefore in line-of-sight with ground-based transmitter and receiver stations for only limited periods, and not on the continuous basis made possible by satellites in geo-stationary orbit.

(iv) Intelsat I (Early Bird), launched 6 April 1965, was the first commercial satellite. It was launched by NASA and operated on behalf of the International Telecommunications Satellite Consortium, Intelsat. Early Bird was sited over the Atlantic Ocean and supplied two hundred and forty telephone circuits, or one TV channel.

(v) The first Molnya 1 was launched by the USSR within a few days of Early Bird, on 23 April 1965. It relayed television, telephone, and telegraph traffic within the Soviet Union. Because of the northerly situation of this country, a different approach to choice of orbit was adopted since a geo-stationary satellite would imply a low elevation from many ground stations. The Molnya programme consisted of a series of satellites maintained in orbit at any one time, all spaced around a highly elliptical orbit. A satellite in such an orbit travels most quickly when nearest the earth (perigee) and more slowly when further away (apogee). Thus the Molnya orbits were arranged so that the satellites reach their apogee over the USSR and hence are in sight of the ground network for several hours at a time. As one satellite approaches the horizon, another appears and the ground stations realign on this, and so on. Unfortunately, the position of the apogee moves in space, and the orbit parameters become unsuitable for lengthy periods, hence new satellites must be launched continuously to replace those no longer of use. This accounts for the large number of satellites which have made up the Molnya 1, 2 and 3 systems.

(vi) The burgeoning application of space satellites in military communications has also been a prime feature. The experiments with Courier in 1960 have already been mentioned, and in 1966 the United States launched a series of operational satellites, IDCSP (Interim Defense Communications Satellite Program). Since that time such programmes as Tascat, DSCS 2 (Defense Satellite Communications System), Marisat, Fleetsatcom, and DSCS 3, for the United States, Skynet 1 and Skynet 2 for Britain, NATO satellites 1, 2 and 3 for the North Atlantic Treaty Organization, and a series of Russian satellites have provided military agencies with global, tactical, naval, aircraft and other operational communications (see UNESCO, 1982).

Communications satellites have often been described as radio relay towers in the sky—like radio transmission towers on earth, only higher. This image might have been relatively adequate at the beginning of the satellite age, but it is no longer. A quick glance at the way in which a satellite in geo-stationary orbit functions might be useful.

Once the satellite has been launched into a geo-stationary track, it is manœuvred into its intended 'parking slot' through small thruster rockets attacked to the body of the spacecraft. The rockets are also used to push the satellite back into its position when it starts to wander off its path, as most do. The satellite is kept pointing in the right direction, i.e. towards earth, by 'spinning': the body of the satellite may spin while the antennas are placed on a de-spun platform always pointing towards earth, or the same effect may be achieved by other methods. The satellite's communications system consists of one or more antennas and a series of transponders—complex electronic devices that receive and transmit signals to and from earth. Since the signals received from earth are weakened by their long journey, the transponders are also used to amplify the signal and convert the frequency, as different frequencies are used for the up-link and the down-link in order to avoid interference between incoming, weak, and re-transmitted, strong signals. Satellites use solar energy; some are covered with solar cells, others are equipped with solar panels, which are continuously being enlarged: with unfolded panel wings, an Intelsat V satellite now used for international communications, is as broad as a five-storey building is tall. The satellite is controlled by a station on earth providing telemetry and command. The signal to the satellite is transmitted from an earth station after the signal has been processed, and the signal from the satellite is picked up by one or several earth stations, passed through a signal-processing system and then injected into the normal terrestrial telecommunications network for delivery to the end-user.

By far the major civilian commercial application of communication satellites to date has been that of long-haul, point-to-point telecommunications links carrying mainly telephone and other traditional telecommunications services, some television traffic and lately also data traffic. Satellites have proved to be a reliable and—particularly on dense traffic routes—cost-effective means of providing these services. These telecommunications satellite systems typically comprise few—though large and costly—earth stations, connected to the terrestrial networks. The recent trend towards more powerful satellites and the gradual introduction of techniques such as digital transmission and single channel per carrier operation permits a reduction in the size of the earth station, so that links carrying less dense traffic also become cost-effective.

The trend towards larger and more powerful satellites can be expected to continue and will result in further reductions in the size and cost of earth stations. This trend can be exemplified through the various types of earth stations operating on the domestic satellite systems of the United States. They have been grouped in three classes:

(i) Heavy route, high quality, multipurpose, sophisticated, large stations costing several million dollars;
(ii) light route, special purpose, smaller stations costing in the $100,000 to $500,000 range;
(iii) broadcast-type, small stations used in cable television systems costing as little as $5,000.

The outlook is for roof-top antennas, for two-way voice or data communication or reception of television only, together with the associated electronic equipment, becoming available in the early 1980s at a cost in hundreds of dollars only, or just simple home-made models.

These developments will enable the use of satellite communications facilities for a large range of applications that would previously not have been economically feasible. Systems currently being developed and implemented will provide, for example: communications networks with individual earth stations situated at each of the different locations within a large business organization; communications with mobile objects, ships, aircraft, trucks; general telecommunications facilities for public use with earth stations close to, or within, each large conurbation in a country and, for remote and isolated communities, however small, the opportunity to have reliable and high quality access to the global networks.

The trend towards smaller and smaller earth stations is foreseen by some satellite protagonists as continuing to the point where direct person-to-person communication via satellite, using 'wristwatch-sized' radio equipment, will be technically feasible. Whether or not this prediction is ever realized, the trend towards smaller and cheaper ground equipment will be a dominant factor in the future development of satellite-based telecommunications systems.

Another significant trend can be seen in the increasing complexity—as distinct from size and power—of satellites. The development of higher-power satellites is accompanied by the need to provide larger total information bandwidth within the radio frequency ranges allocated to different services. Both higher effective power and the ability to re-use the spectrum are provided by higher-gain satellite antennas with small spot-beam coverage. Coverage of the total area is achieved by the use of a large number of individual beams. These satellite developments will result in a loss of flexibility and inter-connection unless special measures are taken to counteract their negative effects and retain their positive features. They are, however, occurring in parallel with the so-called 'digital revolution' through which the means of processing data are becoming both readily available and cheap, and digital transmission techniques are being introduced for the transmission of all types of information.

The next generation of communication satellites will employ digital transmission methods and, by the use of on-board micro-computers and fast-switching techniques, will restore full interconnection between the many separate transmission beams. Such systems will be fully flexible in that any point may be linked to any other point in the coverage area and will also offer the possibility of dynamically allocating the satellite capacity to match the traffic demand.

Such satellites featuring high-power, multibeam antennas and switching facilities will permit the introduction of services that are at present not considered cost-effective; this is due to the relatively low cost of earth stations, the possibility of locating the ground equipment at, or near, user premises, thus excluding the cost of terrestrial links, and the possibility of providing an economical service directly to small or inaccessible communities.

Although transmission costs can be expected to continue on a downward trend, the cost of establishing the space segment of new communications satellites systems will show the opposite tendency, due to the increased power, size and complexity of the satellite and the attendant increase in launch costs. Consequently, the satellite-based telecommunications facilities are likely to come within an affordable range for progressively smaller user communities (or rather user communities having fewer resources) while the establishment of a satellite system will remain the province of large user communities, i.e. international and national agencies. In other words, the technical advances leading to lower telecommunications transmission costs and potentially extending a wide variety of services to a larger user community, will not, *per se*, give the new users any direct influence on the design of the system, nor will they guarantee the availability of the facilities.

Satellite systems and services

At the advent of satellite communication it was generally thought that the use of this new technology would be organized in the form of a few major systems providing all required services in all parts of the world. The present situation is very different from early forecasts; there has been a proliferation of systems at the international, regional and national levels, for a variety of purposes. At the international level, two organizations provide all general telecommunications services via satellite:

— Intelsat (International Telecommunications Satellite Organization), which at present groups some one hundred and six member countries;
— Intersputnik, which includes the USSR, eastern European countries and some countries in Asia, Central America and Africa.

At other levels, the picture is one of ever-expanding use of national and regional satellite systems for point-to-point communications, for example in Canada, the United States, the USSR, Indonesia, western Europe, Japan and—in the near future—the Arab countries and Brazil.

In the case of mobile communications with ships, aircraft, floating platforms etc., navigation by satellite facilities has not received as much attention as other space applications such as telecommunications and earth resources observation. The first satellites to provide navigation services were developed for military purposes but civilian uses were soon found. At present there exists an operational satellite system for maritime communications and new systems are being developed.

The institutional aspects are also under control and an international agency, Inmarsat, has recently been formed to handle communications with ships all over the world.

The use of satellites for aircraft communications is less established. A number of experimental projects have been carried out to provide communications between land-based personnel and aircraft, and more recently a pre-operational system, called Aerosat, was developed by the Europeans and the United States. This has run into a number of political and financial problems and the programme seems to have been stopped for the time being. The twin roots of the difficulty are the cost of re-equipping aircraft and airports and agreement on funding the system, and the fact that, despite criticism and accidents, the current means of communication with, and navigating, aircraft are regarded as being within the bounds of acceptability.

A further form of communications satellite services is represented by the 'direct broadcast satellite' (DBS) which can directly transmit television, radio and other services to individual sets equipped with special antennas; thus the satellite signal can bypass the earth stations connected to the terrestrial network and reach, directly, individual users. The American ATS-6 was an example of an experimental satellite used for early trials of this kind of service in the United States, India and other countries. A more advanced type of experimental broadcast satellite was the Hermes (CTS) system, developed jointly by Canada and the United States. The Japanese are well advanced in this field, with their Yuri system, as are the USSR with their Ekran system. Other systems at various stages of planning and development for such use in Europe include the European Space Agency's L-Sat satellites, the Franco–German bilateral programme for two national systems, the Swedish Tele-X project, the Swiss Tel-Sat proposal, the Luxembourg Lux-Sat plans as well as systems in other parts of the world such as India and the Arab states.

Traditionally, a clear distinction has been made between the fixed-satellite service for point-to-point communication and the broadcasting-satellite service. However, within a very short period, technology has developed to create a quite new situation, in two respects: the distinction between these two satellite services is becoming blurred and furthermore, advanced satellites with broadcasting capability may be used not only for broadcasting but for other purposes as well, such as data transmission.

International regulation

Decisions on the organization and structure of national telecommunications are within the purview of each country. However, for over a hundred years international co-operation and agreement in this area have been important factors in the development and regulation of telecommunications systems and particularly of the international networks.

The ordering and regulation of telecommunications at the international level is developed and agreed within the context of the International Telecommunication Union (ITU) as the specialized agency within the United Nations system responsible for it. The present regulation of telecommunications is a vast and complex field and it is important to understand some of the basic approaches adopted, as reflected in the ITU definitions of activities and services, which differ from those definitions adopted by information or mass media specialists. There is also another reason for emphasizing the importance of the definitions of the different telecommunications services: they provide a basis for general regulation, including the allocation of radio frequencies for radio communication services of which the space services form an ever-growing part.

The general definition of telecommunications focuses on the means and modes of communication that might be used for the transfer of information of any nature as reflected in the definition quoted on p. 27. The definition of radio communication follows logically as 'telecommunications by radio waves', which in turn is divided into two major classes: terrestrial and space radio communications.

Before proceeding to a description of the space services, it should be pointed out that an important definition concerns 'public correspondence', which refers to any telecommunication which telecommunications offices and stations must accept for transmission by reason of their being at the disposal of the public. In practical terms, the backbone of the public telecommunications system is the telephone network, while the other major public system is the broadcasting networks. Alongside these public systems exist private, specialized networks which have increased rapidly with the advent of national and international data transmission.

In keeping with the above definitions, the overall expression used for any kind of space service is 'space radio-communication' which is defined as 'any radio-communication involving the use of one or more reflecting satellites or other objects in space' (Final Acts, World Administrative Radio Conference, Geneva, 1979, Art. N1/1. The first part of the definition refers to active satellites which carry equipment intended to transmit or re-transmit signals. The second part of the definition refers to the passive satellites and to such space objects as the moon, which have also been used for reflecting messages originating from earth back to earth; such reflecting satellites are generally of no further interest in the context of this study.

Following these general definitions, the ITU terminology then defines a series of specific satellite services such as the fixed-satellite, mobile-satellite, broadcasting-satellite, radio-navigation satellite, earth-exploration satellite and other satellite services. The word 'service' also indicates that the satellites themselves are only one component in a system which comprises the transmitting and receiving facilities as well as the equipment for controlling the satellites themselves. In a wider sense, it should also be noted that most satellite systems, particularly those used for general tele-communications including data transmission, are integrated into the general telecommunications networks and thus function within that context.

Also of relevance in this context is the distinction between services that provide communication between two or several defined correspondents with point-to-point communication and those that provide services direct to the general public. To the former category belong the fixed-satellite and the mobile-satellite service, to the latter the broadcasting-satellite service. In 1971, the broadcasting-satellite service was defined as 'a radiocommunication service in which signals transmitted or retransmitted by space stations are intended for direct reception by the general public' (idem). In this service, direct reception encompasses both 'individual reception' by simple domestic installations and 'community reception' by receiving equipment which in some cases may be complex and intended for use 'by a group of the general public at one location; or through a distribution system covering a limited area' (idem).

In general terms, the broadcasting-satellite service was seen in opposition to the fixed-satellite service where messages are transmitted over a satellite system which requires large and expensive earth stations connected to the terrestrial telecommunications network.

The broadcasting-satellite service was thus seen as a distinct service, using heavier, more powerful and more expensive satellites, capable of reaching much smaller and inexpensive receiving installations. However, technology has developed so rapidly that fixed-satellites have been used at least experimentally for services which are almost indistinguishable from the broadcasting-satellite service. As stated by two Canadian experts:

Traditional wisdom is quickly being updated by practical experience. Canadian experience . . . has already demonstrated that 'direct broadcasting' to small earth stations can be achieved with virtually the same low power traditionally associated with 'fixed' satellites. Many countries in addition to Canada, including many developing countries, will be attracted by the flexibility and economy of using the same transponder of a satellite either in the 'broadcasting' or the 'fixed' mode, with the schedule of use of a given transponder being decided during the operational life-time of the satellite system. This flexibility is possible with recent improvements in the design of satellite systems and the realization that from a technical point of view there are such striking similarities between a 'direct broadcast' satellite and a 'fixed' satellite that a single spacecraft can be used effectively in both roles. [Chapman and Warren, 1979.]

Characteristics of communications satellite systems

Conventional wisdom had it that communications satellite systems do not provide any facility that cannot conceptually be provided by traditional terrestrial means, by cable or microwave facilities. Therefore, the satellite would find applications not for a range of services that could not otherwise be provided but in situations where it represents a technically and operationally more flexible and cost-effective solution. Up to the present time these situations have been held to include traffic-dense trans-oceanic routes, domestic systems in countries where difficult terrain or large areas make terrestrial systems too costly, and provision of new types of communications services that could not, within a reasonable time-scale, be offered through extension of existing facilities.

When proposals were first mooted for satellites to carry telecommunications traffic across the Atlantic, there were many who were doubtful about the growth of traffic and hence about the scale of financial returns. They pointed to the existing investment in submarine cables and the greater traffic capacity of satellites was offset against their shorter operating lifespan. The pessimists were, however, proved wrong. Satellites have been extremely successful in both performance and financial terms. Far from obviating the need for cables or other terrestrial facilities, they both complemented and competed with traditional systems and soon became a crucial factor in the emerging new patterns of world-wide communications.

There are four characteristics which together make communications satellites unique. First, they provide bandwidths, i.e. communications capacity, far in excess of those available from conventional terrestrial systems and only comparable to very advanced microwave relay systems and fibre-optic cables.

These new terrestrial facilities do not, however, share the second feature of satellite systems—their flexibility—which is not available in other telecommunications systems. This flexibility manifests itself in a number of ways:

(i) Within the coverage area of any one geo-stationary satellite—theoretically up to two-fifths of the earth's surface—any number of point-to-point links or point-to-multiple-points links can be established through an earth station at each desired location. This is in contrast to terrestrial systems where the end points of every link must be physically connected by cable or linked by a chain of microwave relay stations, each within line-of-sight distance of the next.

(ii) In a satellite system, the satellite functions as a single node or common point while a terrestrial system has to function with many nodes and thus complex routing. Adding new end points to terrestrial systems might therefore involve considerable and often prohibitive cost if the desired end points are distant from the existing network or are separated by difficult terrain such as mountain ranges or water from the nearest end point or node. A satellite system is insensitive both to distance and to intervening physical barriers.

(iii) In complex terrestrial networks, systems growth by extension of facilities and changes in traffic flow patterns affect the required capacity in many links throughout the network and, in addition, often requires considerable up-grading of switching facilities which often represents a major bottleneck. In contrast, a satellite system can be designed to dynamically accommodate any changing traffic flow pattern and any system of growth within the flexible use of the total satellite capacity.

Thirdly, there is an important economic characteristic to be considered in comparing the satellite option with other technologies. Satellite communication is what has been called 'cost-distance insensitive'. In terrestrial systems, the cost of transmitting a message is related to the distance over which terrestrial facilities have been established. In principle, it therefore costs more to connect two telephone subscribers over a longer distance than over a shorter one. Satellite systems do not depend on physical links along the surface of the earth but on the earth stations at the transmitting and receiving points. The costs are therefore related to the use of the satellite and the earth stations: this means that it does not cost more to send a message 8,000 km than 500 km since in both cases the same kind of facilities are used. It would therefore, in principle, have been possible to establish a world-wide unitary tariff for satellite transmissions. However, the establishment of telecommunications tariffs is a complicated business and is based on financial considerations of the costs of the overall network of which satellites form a part. It has been said that the cost savings made possible by satellite technology have not, or not sufficiently, been passed on to the end-user. Even so, it should be noted that international telecommunications tariffs have shown a consistent downward trend in an economic situation where prices for goods and services have shown an equally consistent upward trend.

The combination of these features has enabled satellite systems to take over a major share of the intercontinental telephone traffic and to provide a means for relaying high-information content signals such as television over large distances. It is the fourth feature which provides the opportunity for conceptually new services. By its very nature, a satellite system has a 'broadcast' capability. An earth station can potentially receive all the traffic transmitted via satellite within a given coverage area, or seen from another angle, a single transmitting earth station can communicate with an unlimited number of receiving earth stations, as is done on a smaller scale in a terrestrial broadcasting system. However, the converse is also possible as when a number of transmitting earth stations communicate with a satellite, a data-gathering mode characteristic of services such as meteorology and oceanography.

Most of the actual and planned uses of satellite systems have not been very imaginative in that they mainly represent an extension of existing services, covering larger distances, providing more capacity and flexibility. This has itself been enough to make satellites a key feature in international communications and in domestic communications of an increasing number of countries. Recently, though, new concepts are being studied which would represent radically new communications services and patterns.

It should not come as a surprise that one of these new ideas is under active consideration in India, given the needs of the country, the interest in satellite technology and the proven technical competence. As formulated by

Professor Yash Pal, currently Chief Adviser to the Indian Planning Commission, the proposal is based on a realistic appraisal of the prospects of developing the telephone network in the country:

India is one of the countries with the lowest per capita number of telephones; the total number is less than three million, with inadequate trunk and exchange facilities. If India were to plan for the same number of telephones per capita as obtains in the United States or Canada, say a total of 300 million for a population of about 680 million, the costs would be prohibitive: at the rate of $3,000 per telephone, including the cost of exchanges, transmission lines, the required investment would be about a trillion dollars. This is nearly six times the current gross national product of the country, and is, therefore, a ridiculous proposition. [Pal, 1983.]

Pal and his colleagues envisage a new solution which they have called the 'orbital postman'.

In principle, all the telegraph and telex traffic of the country could be supported by a fraction of the capacity of Insat, using appropriate small ground terminals provided with memory storage through microprocessors. Such a system would be capable of providing a telegraph-type service to all the 600,000 villages of the country without having to go through the bottleneck of metropolitan exchanges. It would not have been possible even ten years ago to consider such a configuration since it incorporates the inexpensive storage devices, computers and other communications techniques that have only recently become available.

In a related but somewhat different proposal, use would be made of a satellite not in a geo-stationary orbit but in a low orbit, at two or three thousand km altitude, which would enable it to circle the earth in 110–120 minutes. If the satellite were provided with a memory storage and a micro-computer loaded with appropriate software, the satellite would be able to convey even long messages between thousands of very simple ground installations. If the satellite is placed in a north–south orbit it will touch all places on earth a few times each day while in an east–west orbit it will cover the tropical countries, most of which also happen to be developing countries. The small ground stations equipped with microprocessors would store recorded messages for different addresses in their memory and transmit them to the satellite in digitally-coded bursts. The satellite would load these messages in its own memory, and travel on, delivering and collecting messages in digitally-coded bursts all around the world. In addition to its cost-effectiveness, such a system would have the merit of being specially tuned to communication-difficult and remote areas; it would have great growth potential and use satellite technology democratically in that all locations would be treated on an equal footing. It would thus meet the needs of telex and electronic mail services within countries lacking adequate communication infrastructure and also meet needs between countries.

Such proposals draw upon and combine in novel ways current satellite-based data collection systems, as used in meteorological and remote-sensing services, new computer information handling and communication techniques. Such new configurations concern developments in satellite technology and ground equipment technology which, together with launcher technology, represent the three corners of a triangle symbolizing satellite communication technology.

The development of satellite technology is thus, on the one hand, dependent on the availability of appropriate launchers that can put communication satellites in orbit, including the powerful and complex, i.e. the heavier launchers of recent and planned design. To a considerable degree, the communications satellite has been a driving force in the development of launcher capacity. A see-sawing process is discernible: originally, only a very few countries had any launch capacity at all but, with time, an increasing number of countries could achieve modest launching capability. With the trend towards heavier payloads the required launcher technology would again be available in some few countries only; should new simpler satellite configurations be implemented, launching capability might yet again be within the purview of a larger number of countries.

On the other hand, satellite technology is linked to the development of ground equipment, i.e. of earth stations for the transmission and/or reception of satellite signals. Here advances in technology have been fully as dramatic as in the other two corners of the triangle. In particular, increasing miniaturization of equipment makes possible new technical systems configurations and new patterns of communication.

The success of the combined launcher-satellite-ground equipment development can be demonstrated by the fact that there are already some 150 communications satellites in orbit around the earth, and some 200 are expected soon. In a very short time we have almost managed to over-use one natural resource which had seemed beyond the terrestrial problem of overcrowding: the geo-stationary orbit. The increasing number of satellites has already created a risk of overcrowding in the preferred positions of the geo-stationary orbit and this in turn must be set in relation to the increasing difficulties of avoiding interference in the signals transmitted to and from adjacent satellites, i.e. a crowding of the frequency bands that according to international agreement are available for space communications. It will, therefore, become increasingly difficult to avoid these problems in a purely regulatory fashion. Consequently, new satellite concepts are being developed or studied by various space organizations and by industry. These include:

— multi-mission satellites which would combine the pay-loads of a number of different operations and put them on one large multi-purpose satellite;
— clusters of satellites located at the same longitude, jointly operated and controlled;
— geo-stationary platforms consisting of various elements that are launched separately into space and then assembled, or making up what in another design version is called an 'orbital antenna farm'. (See Smith, 1979.)

Some observers bemoan the apparent obstacles to these potential developments, which are seen as both desirable and necessary from a technical and economic point of view. These developments do, however, raise the recurrent problem of concentration of 'satellite power' in a few hands in a few countries. In addition, the institutional and legal problems would require new feats of imaginative co-operation as indicated by already existing studies on the subject.

4 The great satellite game: round 1

The sudden eruption of satellites on the communications scene was perceived as a threat as much as an opportunity: a threat to vested interests, to traditional ways of thinking and doing; an opportunity for new world-wide patterns of communication and for commanding these new patterns. Reactions were confused and often oscillated between euphoria and fears of being left out of what might become a major technological and industrial growth area. Governments, idealists and industrialists, military, media and telecommunications professionals struggled for position in what has been likened to a boxing match but should rather be seen as an intricate game. This game, in three if not four dimensions, attracted constantly increasing sets of players, in shifting alliances. The pieces of the game are varied and complex: launching rockets and new international politics, solar cells, computers and the leading edge of high technology, the future of television and the race between the space powers. The stakes were and still are high but the price is worth it: dominion over a key feature of the communications and information world, both of today and of tomorrow.

The American scene

The game starts in the United States. Already in the fifties there had been a growing awareness of the technical and operational advantages of satellites when compared with traditional terrestrial methods of communication, and also of the economic stakes. The potential sources of revenue could be read directly from the statistics of transatlantic telephone traffic which increased by some 15 per cent per year (and at an even higher rate in recent years). The largest American telecommunications enterprise had increased its revenues from international traffic from some $6 million to $42 million in a period of about ten years. Much of this euphoria was reflected in the statement at the establishment of a British space enterprise: 'We who have created this firm believe that there is more money in outer space than anybody could have dreamed . . . Outer space is the Eldorado of the future!'

The players in this first round were drawn both from the private and the public sectors. The private sector players were primarily the conglomerates in the communications and information industry and in the aerospace industry which, endowed with lucrative military contracts, disposed of enough money and talent to construct the satellites and attendant equipment. It is easy to understand the influence and wealth represented by the large aerospace industries but outside the United States itself the power and size of the US communications enterprise are often poorly understood.

In the United States telecommunications and broadcasting have traditionally been owned by private companies under some government control in such matters as frequency allocation, tariff structures and, to some extent, ownership patterns. The most important of these companies, American Telegraph and Telephone, known as A.T. & T., has for a long time been the largest company on earth, with a virtual monopoly over the American internal, and a large part of the international telecommunications traffic. At one stage, A.T. & T.'s investment corresponded to 7.5 per cent of total investment in factories and industrial equipment in the country and its income was higher than that of the five largest states in the USA, which means it was a larger economic unit than most nations. In 1981/82, A.T. & T's position was radically altered, after a decade of vehement combat before the Federal Communications Commission and the courts, arising from new technology, new competing enterprises and anti-trust legal action. Starting in January 1984, the A.T. & T. quasi-monopoly over telecommunications was divested into eight regional companies but, in return, A.T. & T. was allowed to enter into new areas such as computers and data transmission.

The following extract describing the second largest company clearly shows its domestic and external importance:

ITT is a sprawling international conglomerate of 433 separate boards of directors that derives about 60 percent of its income from its significant holdings in at least forty foreign countries. It is the ninth largest industrial corporation in the world in size of workforce. In addition to its sale of electronic equipment to foreign governments, and operation of foreign countries' telephone systems, roughly half of its domestic income comes from U.S. Government defense and space contracts. But it is also in the business of consumer finance, life insurance, investment funds, small loan companies, car rentals and book publishing. [Johnson, 1968, p. 44.]

The same author speaks of the American communications scene as being subject to 'local monopolies, regional baronies, nationwide empires and corporate conglomerates'. This, then, is the private sector which confronted the public sector at the advent of satellite communications. At the public service top, apart from Congress and its various committees concerned with communications and with space, there were, first of all, NASA and the various branches of the armed forces. Also, within the Executive Office of the President, a special co-ordinating body had been created, then under the title of 'Office of Telecommunications Policy'. There were also many agencies in the luxuriant flora of American administrative bodies that directly or indirectly were, or wanted to be, involved.

Of particular importance was one of the more controversial flowers in this administrative flora: the Federal Communications Commission, the FCC. The FCC had been created in 1934 to oversee and order the entire telecommunications sector which in American parlance included the broadcasting industry. Its mandate ranged from frequency assignments within the country, via supervision of property rights in broadcasting to the laying down of tariff rules—to which was now added control over the establishment of communications satellite systems. Although it seemed practical to establish a body responsible for co-ordination of the entire telecommunicatons sector, the FCC has proved controversial, to say the least. Many are of the opinion that the FCC has not fulfilled its tasks in accordance with the Communications

Act of 1934 but rather has become a pawn in the games conducted by the strong interests and entities in the communications and information industry.

The first round in the great satellite game concerned the question of whether space communication is more space or more communication. Ridiculous question? Hardly, when we are dealing with future markets of hundreds of millions of dollars, with old and new positions of power—all this tied to the rights of ownership and operation of satellite systems. During the year 1961 no less than twelve congressional committees studied the real bone of contention: the rights of ownership.

Despite the fact that it was the taxpayer who had footed the bill for American space activities, President Eisenhower had managed in his farewell speech to mention the constructive and eminently American role that free enterprise could play in outer space, while in the same breath he warned his countrymen of the military–industrial complex!

Since NASA's research and development programme in this area was mobilized around the objective of promoting operational satellite communications systems, the question of technology transfer became a major issue. The rapid development of the technology required a policy in advance of actual satellite experiments and demonstrations. Significant factors involved in the implementation of such a technology transfer policy were:

  (i) the existence of a commercial market for communications satellites;
 (ii) the American decision to be first with an operational system, especially with reference to US–USSR rivalries and competition, and
(iii) a traditional approach to government-sponsored research and development which results in eventual transfer to private sector control of potentially revenue-bearing activities.

Following a number of inquiries and studies, various approaches to the establishment of a satellite system for international traffic were discussed. The FCC proposed a non profit-making management corporation to develop and manage the satellite system on behalf of the common carriers (i.e. the private telecommunications companies) which would own both satellites and earth stations. This attitude met with widespread opposition. A public committee considered the three basic options which seemed to be open: government ownership, private ownership by the carriers, and broad-based private ownership. In one appraisal it is said that,

these options were not examined according to their technical and organizational consequences alone, but of necessity were filtered through the political and psychological climate of opinion. This ethos concerned certain vague, but highly emotional issues such as 'creeping socialism' and 'predatory monopolies' . . . In short, the issue was political: it involved polemics and emotions as well as dispassionate consideration of the pros and cons. [Galloway, 197–, p. xx.]

In 1962 the Administration did finally put forward a bill to Congress but there were, in addition, a total of fifteen legislative proposals submitted for implementing communications satellite operations. In principle, the proposals reflected the three options concerning the ownership mentioned above; the Administration Bill proposed broad-based private ownership, another bill favoured private ownership by existing carriers and a third bill advocated government ownership.

Feelings ran sufficiently high for attempts to be made in the Senate to begin a filibuster to block the passage of the Administration Bill. However, these attempts were defeated and in August both houses passed the bill which was signed by President Kennedy into law on 31 August 1962.

The Communications Satellite Act

The Communications Satellite Act has been judged in very different ways. In one version,

the creation of a public/private corporation was unique. Although the concept of statutory-created corporations was not new, the dual public/private nature of the Communications Satellite Corporation was unparalleled in scope. The Communications Satellite Act created a private corporation and granted it a monopoly in the business of intercontinental satellite communication for the expressed purpose of facilitating the public goal of developing a global communication satellite system. The corporation would serve both private and public interests. The Comsat Act established an elaborate system of checks and balances to ensure the public/private character of the enterprise. [Smith, 1976, p. 108.]

Other observers were of the opinion that the Act was one of the great errors of the Kennedy Administration. One of the harshest critics was former President Truman who said that he did not think that the President (Kennedy) understood the bill.

The damned republicans and some democrats try to give away public property . . . there can be no justification for giving this vast resource that has been financed by the tax payers away to a small group of stockholders for their private gain. The tax payers have already paid for their right to share in the returns. [Goulden, 1968, p. 44.]

The Communications Satellite Act states specifically that Comsat (Communications Satellite Corporation) would not be an agency of the US government. The government would neither own any portion of the Corporation nor guarantee investments to ensure the profitability of the venture. The 'Comsat Act' was designed to protect the public interest by establishing a structure of internal and external controls over the corporation management. Internal controls were supposed to be provided through a broad base of ownership and through government representation on the board of directors. External controls were established through legislative provisions which specify the Federal Government's range of authority to co-ordinate, plan and regulate Comsat activities. The president, the FCC, NASA and the State Department were all given certain areas of authority. The corporation was to be financed through the issue of common stock of which 50 per cent was originally reserved for purchase by the common carriers.

The Act had, however, avoided some difficult questions, and the Senate had very soon to grapple with the problem of the respective mandates of NASA and Comsat in the communications field. In connection with the NASA budget appropriations, questions were raised as to whether taxpayers, through NASA, should underwrite any research benefiting the stockholders of the profit-making Comsat; another bone of contention concerned the reimbursement by Comsat for satellite launchings by NASA. Also, nothing was said in the Act concerning domestic satellite communication. Further difficulties were caused by the different and contradictory roles some organizations played in relation to Comsat. Thus, A.T. & T. was for some time the largest individual stockholder, with a seat on the board of directors. Through its ownership of transatlantic cables, A.T. & T. was, simultaneously, a competitor; through its lease of satellite channels, also a client; through its manufacturing activities, also a purveyor—and finally, through its ownership of the terrestrial transmission network its co-operation for any satellite transmission was required.

The next round in the great satellite game concerned the ownership of the earth stations, which are decisive for the revenue to be derived from satellite communications. The cost of the space segment represents only one part—and not the most expensive part—of a satellite transmission. The costs of the land lines to and from the transmitting and the receiving earth stations and the costs of the use of the stations represent the major portion of the costs for a satellite transmission. Comsat and the common carriers both wished to acquire this plum. Following rather heated discussions, the FCC decided that the American earth stations would belong in equal parts to Comsat and to the common carriers and that the revenue would be divided accordingly. A.T. & T. and the other carriers could thus add one further role in their relationship to Comsat.

International aspects

The following round of the game was more international in character. The two major objectives of United States foreign policy on satellite communications were to protect the national interest and to further international co-operation under American leadership. Early international contacts had mainly been conducted by NASA with the telecommunications agencies of other countries, in particular those of western Europe, Japan and Canada. As approved by Congress, the Communications Satellites Act of 1962 made the United

States dependent upon the uniquely structured public/ private Comsat Corporation for the implementation of its communications satellite goals. Congress had established Comsat as the chosen instrument of American policy in the belief that private enterprise, with its acknowledged managerial skills, would be able to develop a global system in an efficient and expeditious manner. The actual development of an international system would necessarily depend on the interest and involvement of other nations. In this respect there were, from the beginning, conflicts between Comsat and the State Department. Comsat originally expected a series of bilateral agreements while the State Department argued for a multilateral negotiating process. And again the question of ownership cropped up: Comsat wanted to own and manage the international system itself while the State Department argued for a truly multilateral international organization that would own and operate the system.

While the United States was engaged in clarifying its own policy other countries began to formulate the principles that would govern their negotiations with the States. Seen from the outside, Canada and Japan presented an image of a relatively well co-ordinated policy process. The situation in Europe was otherwise confused. Individual countries had great difficulties in evolving a coherent policy and even the institutional structure for dealing with satellite communications. Some countries created new institutions, such as Italy where a new body, Telespazio, was entrusted with a major policy-making and operational role. In other countries, satellite communications were, often by default, located in the telecommunications administrations, which led to problems with other concerned agencies. The lack of coherence at the national level was mirrored at the regional level through the existence of a number of European space organizations with different goals, principles and membership. Finally, the European countries tried to co-ordinate their approaches through a new organization, the European Conference on Satellite Communications—generally referred to as the CETS, the acronym of the French title, Conférence Européenne des Télécommunications Spatiales. It has been said that European solidarity, however limited, served as a catalyst to bring about agreement between Comsat and other members of the American *ad hoc* Satellite Group. Essentially, Comsat and the government policy-makers decided to press for international ownership and multilateral negotiations. While the United States was prepared to begin the talks it was still necessary to persuade European and other nations to enter into formal discussions. According to NASA, other nations were dragging their feet because they feared that the United States represented by Comsat would dominate the system. Thus they were delaying the commencement of negotiations in order to gain a more effective negotiating position.

Intelsat

However, negotiations did start in the autumn of 1963, with the participation of the western European countries, Australia, Canada and Japan. The fact that there were no developing countries involved reflected what was called a 'pragmatic approach' since more than eighty per cent of the world's international telephone traffic was conducted by the western industrialized countries.

From the American side great emphasis was laid on the fact that the negotiations should concern one single global —and for all countries—open commercial system. And the United States was in a hurry. It was therefore expedient to concentrate on provisional agreements and postpone the decision on the definitive forms of organization.

The organizational structure that was finally adopted in 1964 was unique in inter-governmental co-operation; a majority of participating countries were of the opinion that it could only be accepted on an interim basis. This was reflected in the title of the main agreement—'Agreement Establishing Interim Arrangements for a Global Commercial Communications Satellite System'. This agreement, which created the International Telecommunications Satellite Consortium, Intelsat, was signed by participating governments and included general organizational and administrative provisions. It was complemented by a 'special agreement' at the insistence of the Americans who wanted to provide for an arrangement whereby a formally non-governmental organization like Comsat could enter as partner into these international contractual arrangements. The special agreement could therefore be signed by recognized private agencies such as Comsat.

The agreements concerned the space segment of the satellite system, including the earth stations for telemetry and control of the system. The earth stations required for the transmission and reception of signals should follow certain technical criteria defined by Intelsat but they were to belong to the countries that established and operated them and not to the consortium.

The capital investment, originally $200 million, was divided among participating countries in relation to their part of the international telecommunications traffic. The somewhat controversial methods for calculating these parts resulted in the quota for Comsat, i.e. of the United States, being 61 per cent; 30.5 per cent went to the western European countries and the rest to Australia, Canada and Japan. These quotas would be changed with the adherence of further countries but in no case would the United States' share drop below 50.6 per cent. The system was pertinent in that voting strength was linked to the investment, as was the membership on the Interim Communications Satellite Committee, i.e. they provided for a unilateral American right of veto. The voting rules were complemented with a set of complex regulations which were designed to limit somewhat the absolute power of Comsat

As could be expected, Comsat held the dominant position. In its following annual report to shareholders Comsat mentioned, with apparent pride, the three roles it played in relation to Intelsat: as member, as the US representative— and as executive body. Thus, as an American agency, Comsat was under the direction of its own board, of the FCC and other American authorities, while as executive agent it should only pay attention to common international concerns. It would have been impossible for any organization to hold such a split loyalty.

During the years following the interim agreements, Comsat and its foreign partners quarrelled frequently, despite the fact that a system was established and performed well from the technical and commercial point of view. Under the direction of Comsat, the consortium was launching one generation of satellites after the other and the entire system was producing substantial revenue. Yet many of the areas of disagreement that characterized the 1964 discussions began to reappear during the second half of the 1960s.

These internal weaknesses were complemented by an external weakness: the interim Intelsat agreement could in principle be interpreted in two different ways. On the one hand, it could be seen as an attempt to implement the principles of a UN resolution which recommended the early establishment of communications satellite systems available to all states without discrimination. On the other hand, it could and was seen as an attempt to circumvent the UN and to counteract the intention of the resolution, since the rules heavily favoured certain countries only, and mainly the United States. One major problem from the outset was the word 'single', in the phrase 'a single global system'. The manner in which the voting rules had been defined made participation by the USSR and certain other countries impossible. Consequently, the idea of a single global system sounded exaggerated, naive or false since one group of countries cannot bind other countries through a decision in which they have not participated.

These and still other contentious issues influenced the negotiations over the definitive Intelsat agreements that started in Washington in February 1969. At the outset American authorities expected to finish the conference in some four weeks. Because of the deep-rooted differences of opinion on all basic questions it took instead two years to reach results—and these were a compromise that seemed to wholly satisfy no one.

While there were differences of opinion between the United States and the European countries on such matters as American control of the transfer of technology and the placement of hardware contracts, there were also differences in policy between the industrialized countries and the developing countries that now took an active part in the negotiations. Even more serious were the questions of the scope of Intelsat's mandate in relation to the establishment of other satellite systems and the perennial controversy over Comsat's dominant position in the organization.

The US policy-makers believed that the existence of separate or regional systems would undermine the global efforts of Intelsat—and thereby also undermine the dominant American position. In fact there could be two major reasons for launching a regional or domestic satellite system: nations large enough to support such a system wanted to retain control over their communications and nations also wanted to gain competence in the development of communications and aerospace technologies. Indeed Eurospace, a consortium of European space industries, advocated the establishment of a regional system simply in order to improve the overall capabilities of European industry. Canada had already, in 1968, decided to establish a domestic system and at the same time NASA negotiated for a large-scale experiment in India using an American satellite with a view to preparing for a national Indian system. Also, the fact that the United States had proposed its own domestic system as early as 1965 made it difficult to maintain opposition to separate systems. Even so, this is an issue that is still very much alive.

The growing dissatisfaction of the European countries also resulted in sharp criticism of Comsat's dominant position in the consortium. It was realized that the United States could maintain control 'only at excessive political cost' (Smith, 1976, p. 144). The existing formula for allocating votes according to the national use of the Intelsat system was another primary factor in the growing pressure for independent systems.

The Europeans had, as usual, difficulties in arriving at a consensus. However, they did manage to agree on the desirability of removing Comsat as the systems manager, restricting its voting power and maintaining the right to develop, on certain conditions, separate or regional systems.

Negotiations continued into 1969 and for the following two years. The early American proposals proved unacceptable to other members of Intelsat, many of whom presented proposals of their own. The final Plenipotentiary Conference was held from 14 April to 21 May 1971 and was able to finalize the text of the definitive agreements. In contrast to the 1964 agreements, the definitive arrangements marked an important shift from exclusively technical and operational considerations to a mixture of these, political, and institutional concerns.

Intelsat is established under two international agreements. The first one is concluded between governments which are 'Parties' to the Intelsat Agreement. The second one, known as the 'Operating Agreement', is concluded among 'Signatories', i.e. the telecommunications or similar entities designated by their governments as the investors/owners/operators/users of the Intelsat system. In order to meet the demands from smaller user nations for a greater voice in the decision-making process, the previous consortium was reconstructed into four distinct organs, two of which—the Assembly of Parties and the Meeting of Signatories—offer general membership and decision-making powers on a one-nation/one-vote basis. A third administrative organ, the Board of Governors, composed of Signatories, would be responsible for the 'design, development, construction, establishment, operation and maintenance of the Intelsat space segment' and for the adoption of plans and programmes 'in connection with any other activities which Intelsat is authorized to undertake.' The board was to be composed of up to twenty-seven members, the voting strength of each individual member being defined in proportion to its use of the satellite system. As a concession to concern over the dominant position of Comsat, the agreement stipulates that an individual governor cannot cast more than 40 per cent of the total voting power. Concerning management activities, which represented the fourth organizational level, agreement was reached on the eventual replacement of Comsat as systems manager by a director general. Thus, the definitive arrangements inaugurated a new era for Intselsat, now officially known as the International Telecommunications Satellite Organization.

The first major decision made by Intelsat was the establishment of a global satellite system using geo-stationary satellites. The first communication satellite for commercial operation, Early Bird (Intelsat I) had been launched on 6 April 1965 and placed in orbit over the Atlantic. Two years later followed the Intelsat II series: one was placed over the Atlantic Ocean and two over the Pacific. At the end of 1968 and early 1969 satellites of the following generation were launched. The Intelsat III satellites had a much greater capacity than the earlier types. They were placed in orbits over the Atlantic, the Pacific and the Indian Oceans. In this way almost every part of the world could be reached by satellite.

All subsequent generations of satellites have been larger and provide greater capacity. From the single small Early Bird of 1965 which weighed a mere 85 pounds and had a capacity of no more than 240 telephone circuits or, alternatively, one television channel, the organization is now planning to launch, in 1984, six modified Intelsat V satellites, each with a capacity of 15,000 telephone circuits, plus two wideband television channels. And at the same time the cost of the use of the Intelsat space segment decreased dramatically: 'Indeed, when adjustment is made for global inflationary trends, the cost of a full-time Intelsat circuit ($9,360 per year) is in effect 18 times less than the cost of service 18 years ago' (Pelton et al., 1983, p. 21).

At the same time, the number of members in the organization increased, as did the number of earth stations working with the system. The original eighteen members soon reached eighty and it is now well beyond one hundred. At the end of 1965 there were earth stations in five countries (the United States, France, Britain, West Germany and Italy). Already by the end of 1972 there were 74 earth stations in 50 countries, and in 1982 some 300 earth stations in some 146 countries.

Molnya and Intersputnik

However, the next round in the great satellite game was not late in coming. For some years the Americans had been the only ones to have constructed satellites for communication purposes although the Russians had also discussed the advantages of communicating via satellite. In statements since the early 1960s there had been mention of passive satellites as well as very large and heavy, active satellites. Although it seems that less has been published in the Soviet Union on satellite communications, the internationally-known Academician Blagonravov had, in 1961, pointed up the importance of the use of sputniks for radio communications. He specifically mentioned the need to learn how to use new frequency ranges and to develop instruments that could use inter-molecular and inter-atomic fluctuations.

The prime position held by the Americans changed when, on 14 October 1965, the Russians launched their first communications satellite, type Molnya. The Molnya satellites were heavier and larger than the American satellites and were used for many purposes at one time: telecommunications, television transmissions and meteorology. In contrast to the American-developed satellites, which were intended for international communication, the Molnya system was primarily intended for domestic use.

The Russian satellites were also so powerful that the earth stations could be smaller and cheaper than those used in the Intelsat system, despite the fact that they had to include tracking facilities to follow the non geo-stationary Molnya satellites. The system functioned in a distribution mode and was largely used to transmit television programmes from Moscow to regional receiving stations from where the programmes were distributed by normal terrestrial means.

The entire system, under the name 'Orbita', was officially inaugurated at the Fiftieth Anniversary of the October Revolution, in 1967. The ceremonies at Red Square were transmitted to some twenty stations—one as far away as Vladivostok.

Already in May 1967 a number of socialist countries had, at a meeting in Moscow, decided to co-operate in the exploration and use of outer space; specific mention had been made of communication satellites. Total silence reigned over these plans until these countries, under the leadership of the USSR, and shortly before the UN Outer Space Conference in Vienna in August 1968, publicized a proposal to establish a new international system for 'communications using artificial earth satellites', under the name 'Intersputnik'. The USSR contacted some forty countries where it expected to find some interest but the system was in fact established by the nine original socialist countries: the USSR, Poland, Hungary, Bulgaria, Czechoslovakia, East Germany, Mongolia, Rumania and Cuba.

There are a number of similarities between Intersputnik and Intelsat but also differences. In both cases the intention is to establish an international satellite system even though Intelsat speaks of a global system. In principle, the space segment including control stations is in both systems the common property of the organization while the earth stations are owned by the participating countries. Even non-member countries can use the system, on certain conditions. And while in Intelsat one country has an explicitly dominant position by virtue of its voting quota and operational rules, so, in Intersputnik, similar rules are valid for the payment of satellites, for launches and for the control system. The key to the system is in both cases firmly in the hands of the space powers.

Intersputnik and Intelsat differ primarily in organizational and legal provisions. In the introduction to the Intersputnik agreement, reference is made to the respect for the sovereignty of countries which, in concrete terms, is supposed to mean 'one country/one vote'. Intersputnik was to have a board on which all member countries are represented. Decisions were made by two-thirds majority. While Intelsat was open to all states that were members of the ITU, Intersputnik was open to 'all countries'—which at that time was a reference to China, North Vietnam, North Korea and East Germany. Within the Intersputnik system communication channels were to be allocated on the basis of the needs of various countries for telephony, telegraphy and the exchange of television programmes. The financing of the Intersputnik system should be carried out on the basis of contributions from countries corresponding to the use made of it by the members.

Intersputnik was inevitably seen as a challenge to Intelsat. Many observers were of the opinion that the Russian plan had contributed to make a difficult situation even more difficult. Others thought that the USSR initiative indirectly served the purpose to make necessary new agreements over a global system. In any case, the question remained open, where all other countries would find themselves in this contest between the space powers.

In the meantime the great satellite game has become ever

more complex through the advent of new pieces and more players. So far we have discussed a game which, at least according to current definitions, was civil and peaceful. But in the same way that space activities are a mix of civilian and military uses, so also can almost any kind of space communication serve either civilian or military purposes. For example, the meteorological data from a satellite are equally useful for civilian as for military navigation. There is not much that an observation satellite cannot see, whether cloud formation, mineral deposits, or military installations. Science and spying are so interconnected that they can only be separated by the intention of those involved. As with any other large-scale undertaking, the military cannot function without communications, and satellites have provided a powerful new tool.

Characteristically, the military of both superpowers quickly grapsed the importance of this new means of communication. In the United States, true to their traditional competitiveness, all three defence forces, the Army, the Navy and the Air-force, insisted on having their own space and satellite programmes. It is beyond the scope of this study to enter into details—but it is clear that, early on, the Americans established a number of military communications systems; that one system has been established for use by NATO and that a number of the Russian communications satellites had a military function. In fact, as mentioned earlier, the use of satellites has become such an essential feature of military planning that the mere possibility of threats to military satellite communications—in the form of satellite-killer weapons—has become one of the major causes of the dangerously increasing militarization of outer space.

5 Uses and users of satellite communications

For the general public the most obvious uses of communication satellites have been the television spectaculars: the Olympic Games, assassinations, papal visits and royal weddings have been transmitted live to national television stations for retransmission to audiences which by now are counted in hundreds of millions. But television only represents the tip of an iceberg. By far the heaviest use has been for telecommunication purposes with national telecommunication administrations (PTTs) and private common carriers as the most important users, providing satellite services to their clients, the end-users, within the framework of the overall, general telecommunications networks which they operate. Recently, however, new patterns have begun to emerge: satellite services are offered directly to the end-user who then deals directly with the providers of satellite services. How do satellite services appear to the user in these different cases? What possibilities have the users to influence the design of services to suit their purposes? What are the constraints and the opportunities?

Services and users

From the user viewpoint, satellite telecommunications services could be provided in the form of:

(i) one or more complete satellites dedicated to a single user;

(ii) a part (e.g. single transponder or channel) of one or more satellites dedicated to the user with other parts dedicated to other users;

(iii) a satellite serving the needs of several different users with part of the capacity (but not a physical part of the satellite) leased to each;

(iv) the 'incidental' use of a satellite link as a part of general service provided by a common carrier, PTT, or similar organization. (See UNESCO, 1982, p. 43.)

At present, the largest use is represented by case (iv). Examples of this are a telephone subscriber making an intercontinental telephone call, or the leasing of a private circuit in a country where the service can be provided by a domestic satellite system. In such instances the satellite communications facilities will be provided by the telecommunications agencies as an adjunct to other means of transmission. The end-user will generally not know and not particularly care which facility is used as long as the messages come through rapidly, efficiently, and as cheaply as possible.

This use of satellites has so far had little impact on the user. It has neither significantly increased the range of services available, nor has it had any major deleterious effect on the quality of other technical characteristics of the services. One exception concerns the geographical patterns of international telephone traffic. Traditional networks were established so as to connect the major colonial and metropolitan centres with the rest of the world. Thus, it was easier to call London or Paris from an African capital than places in a neighbouring country. Similarly, telephone calls from Buenos Aires to Santiago had to pass through New York—at the expense of the user. Satellites have completely changed all this, since the connection does not depend on the physical emplacement of terrestrial facilities: any connection is possible between countries possessing earth stations linked to the international satellite systems. And even though all the cost savings of satellite communications have, for a variety of reasons, not been passed on to the end-user, the tariffs for intercontinental telephone, telex and data connections have decreased considerably since the introduction of satellite facilities.

Of more interest are services provided under categories (ii) and (iii). Within limits imposed to protect other users of the satellite and geo-stationary orbit, the users in these categories will have some freedom to influence the architecture of the system and select the most appropriate operating parameters. Obviously a user leasing one or more complete transponders would have considerably more freedom in this respect than a user only leasing capacity on a satellite.

Examples of various users in these categories include, for category (ii), the leasing by the Norwegian government of an Intelsat transponder to provide a communications service to the North Sea offshore oil installations. Here the satellite transponder characteristics are designed for the normal Intelsat telephony traffic, but some aspects of the earth terminals and the system organization are specific to the service. For category (iii) an example would be a business organization renting capacity on a satellite such as the American SBS system for a computer-to-computer link.

In any system where various users share the same satellite there is no reason why the different services should be of the same type. It is, for example, technically feasible for a single satellite to simultaneously provide a public broadcast service and various telecommunications services or meteorological services. Indeed, both technical and economic reasons seem to push in the direction of such multi-purpose satellites. However, from the regulatory side, there is now a tendency, which was evident in the deliberations of the 1979 ITU World Administrative Radio Conference, to allocate different sectors of the geo-stationary orbit to different classes of services. Such regulatory constraint will thus prevent such sharing between certain, though not all, classes of service.

An important distinction between categories (ii) and (iii) must be drawn. In category (ii), where whole transponders are allocated, economic considerations dictate that the user's volume of communication needs must be commensurate with the typical capacity of a satellite transponder. This capacity is of course dependent on many parameters, including the characteristic parameters of the earth stations to be employed, but as an indication the capacity of a single transponder will permit a throughput of many tens of megabits per second (Mb/s) of data.

If a user can justify the generation of these relatively large volumes of traffic and has the resources to implement a large system of this type, then the freedom to optimize the system for the exact traffic characteristics can yield significant economies in transmission costs. For example, the allocation of capacity to the various links need not be 'balanced' (equal flow in both directions), resulting in less wastage of system capacity where the traffic flows are unbalanced. On the other hand, the use of category (iii) implies that the provider of the capacity will specify a system which is a compromise for many users and, while permitting smaller users to participate, might result in higher transmission costs.

Both users and their requirements vary with the geographical levels at which information transfer via satellite can take place. In this respect, two main levels are relevant: information transfer between countries and information transfer within countries. For long-haul international transmissions a given country would normally choose to be connected to one of the international satellite systems now in operation. At present only two such systems for general telecommunications are in existence—Intelsat and Intersputnik. While the availability of transmitting and receiving facilities in a large number of countries is of advantage for the use of international satellite systems in information transfer, the present system also presents considerable disadvantages.

The two international satellite systems are designed in such a way that they tend to favour areas with an existing, developed telecommunications network and with heavy traffic requirements, i.e. mainly large urban areas. The power of the satellites is relatively weak, which means that the associated earth stations have to be relatively large and expensive. This in turn means that the system configuration provides for one earth station—or in the case of very large countries, several stations—connected to the terrestrial telecommunications network. The system is therefore not designed to provide links between numerous points within countries to similarly dispersed points in other countries. It should be pointed out, however, that with the advance in technology there are possibilities for using small earth stations which might in a number of cases represent a more adequate system configuration for small and dispersed users, and particularly for rural telecommunications in developing countries.

A number of countries have found it advantageous to use satellites for internal communications. Such national use of satellite facilities comprises all forms of telecommunications, from telephony to television and data transmissions. In terms of general policy, the options available are mainly three: a country (i.e. the relevant telecommunication authorities or in certain cases other operating agencies) may lease facilities on an existing international or regional satellite system; it may participate in the establishment of a regional system; or it may establish a satellite system of its own.

Obviously only a few countries will require, or be in a position to afford, a satellite system of their own. Among the countries that at present have set up, or are in the process of setting up, domestic satellite systems are, as could be expected, the larger industrialized countries: Canada, France, Germany, Japan, the United States and the USSR. However, the satellite option is in certain circumstances technically and financially attractive enough for developing countries to have also chosen this technology. Indonesia has now for some years operated its own Palapa system and decisions on setting up national systems have been made by such countries as Brazil, India and others.

Other countries, both industrialized and developing, have so far chosen to lease facilities available on the Intelsat system (e.g. Algeria, Malaysia, Nigeria, Norway, and others). Another possibility is to lease facilities available on a domestic system; thus Malaysia, the Philippines and Thailand have leased channels on the Indonesian Palapa system. This option may obviously present a number of constraints of a technical, operational and legal nature and is in principle open only to countries whose territory can be adequately reached by the beams of a neighbouring country's satellite.

However, all these projects would still not solve the major problem in current communications development: the grossly uneven distribution of basic telecommunications facilities between and within countries. At present about 15 per cent of the world's countries own 90 per cent of its telephones; within developing countries most telephones are found in the cities and are owned by the elite. Although organizations such as the World Bank have given telecommunications a low priority, there is now a belated recognition of the socio-economic benefits of telecommunications development. A sign of this change in attitude is represented by the new revolving fund for telecommunications development being set up by some twenty-two industrialized countries; another is the ITU plan for a global satellite system designed to provide domestic telephone services in rural areas. The aim of this plan is to put a public telephone within one hour's walk of each settlement and village in the world. This coverage would be achieved through a new world satellite system estimated to cost about $300 million, plus earth stations costing $15,000 each.

Since, in principle, there are as many satellite communications services as there are telecommunications services, it will be more interesting and revealing to look at some specific examples of opportunities and problems. The choice has fallen on the transmission of information in two distinct forms: as news and as data and then on the educational uses of satellites as one of the more controversial aspects of satellite communications.

Information exchange: news

Gertrude Stein's famous saying 'a rose is a rose is a rose . . .' has been given many interpretations: one is simply to say that a rose is irreducible and self-defining. The same seems to be true of 'news'. If one listens to the explanations of news practitioners—journalists in print and in the electronic media—one is left with the idea that 'news is news is news . . .', no more, no less. To use the term 'information' does not help much: there is no agreed definition of information; rather, each profession—be it journalism, computer science, librarianship or the law—has its own definition. And moreover, all information is not necessarily news—however, in this perspective it is easy to ascertain at least one characteristic of 'news', as generally used: news should be made available as fast as possible to the public—however this 'public' is defined. The history of news can therefore from this standpoint be correlated with the history of transmission. The purveyors of news have been among the first to use any new means of transmitting and distributing information, by telegraph, telephone or telex, by wire, aircraft, wireless—or satellite.

Technically, the press, the news agencies and radio could adequately be served by traditional telecommunications facilities. Not so television, which demands so much more bandwidth, i.e. transmission capacity, for its moving images that the costs of investment and operation of the facilities required are of another order of magnitude. So that, when television started, there was no alternative to sending news films by air, which might mean days of delay. The result was that early television news was often no more than 'talking heads' with the addition of the odd slide or map.

If television were to be taken seriously as a news medium, more was obviously needed: pictures of actual events, live or recorded—but not days after the event. The impetus for international electronic transmission of television news via terrestrial systems began in western Europe. In the second half of the 1950s the west European broadcasting organizations managed to include the transmission of news in the activities carried out within the framework of Eurovision. Eurovision stood for and still stands for a transmission network and a framework—organizational, financial, legal— for the exchange and transmission of television programmes and materials under the auspices of the European Broadcasting Union. The problems were numerous and often caused strain between the broadcasters and their national PTTs: the need for adequate transmission facilities, conditions of access and use of the landlines, and last but not least, the tariffs to be paid, which broadcasters have always found too high.

Despite these difficulties, sufficient agreement was reached for news transmission to rapidly become one of the most important Eurovision activities. Even the arrival of news films from other continents could be speeded up by injecting them into the network from the capital best located in terms of airplane schedules: London for news film from North America, Rome for films flown in from the east and the south. Thus, flight schedules and not editorial judgement to a large extent decided what the public was offered in the main evening television bulletin.

This situation changed with the advent of satellites as the first transmission facility with enough capacity to transmit moving pictures over intercontinental distances. As soon as satellites arrived on the communications scene, their importance in the transmission of news for television was therefore well recognized. There were great expectations—and even greater problems. The story of the attempts to improve the dissemination and exchange of news via satellites is long and complex.

Such technical difficulties as the differences in television standards between North America and Europe and within Europe had to be overcome, through the development of special equipment for the conversion of signals. But conditions of access, operational procedures and tariffs became ever more important objects of negotiation between the broadcasters, the telecommunication administrators and the operators of satellite facilities. In these negotiations the PTTs often wore several hats, as compared to the broadcaster's one: in most countries they owned the telecommunications facilities required to get the signals from and to the earth stations; in Europe, they also owned and operated the earth stations and they were partners in Intelsat which owned the space segment.

It is not the intention to chronicle here the endless series of consultations, meetings and negotiations over the conditions for use of satellites for news transmission, nor to enumerate the equally endless number of committees, authorities, groups and agencies involved. Apart from conditions of access the main problem concerned the high charges levied by the telecommunications agencies which caused problems for all broadcasters, but particularly those in developing countries.

As pointed out earlier, the division of responsibility in the communications and information field is in most countries allocated in a manner which has become inadequate. In general, telecommunications are handled by public or semi-public telecommunications administrations (PTT) or by commercial enterprises (common carriers). Commercial bodies are obviously supposed to make a profit on their activities but even the public administrations are in most countries enjoined to be at least self-supporting, hopefully also to make a profit for the national treasury. Broadcasting, which is both a telecommunications service and a journalistic medium, generally falls under the authorities dealing with information. From the information side, the requirement is for technical facilities to be available cheaply to assist broadcasters in the collection and distribution of news, while from the telecommunications side the requirement is to run a profitable operation. The contradiction between these two requirements can only be solved at the policy level, but generally and until recently, tariff questions have been regarded as unworthy of the attention of policy-makers and have been left to unilateral decisions by the telecommunication agencies, under some government supervision in certain countries. The result was not only to hamper the flow of news between countries but also, indirectly, to channel this flow. For some years the European telecommunications agencies had charged tariffs for the use of earth stations which were considerably higher than those in North and South America. Thus, it was cheaper to route news material from South America to New York, to edit it in New York and then transmit it to Europe. This meant that the European public was treated to information on events in South America seen from an American perspective, which raised serious questions with regard to editorial judgement, cultural and political perspectives.

These questions were put into a wider context with mounting demands by the developing countries for a new world information and communications order. Ministers of information and even heads of state were clamouring for a more balanced flow of news, particularly between developing countries themselves, and from developing to industrialized countries. It was recognized, albeit late in the day, that high satellite system tariffs were one obstacle to the extension of news flow. However, for years there had been a tendency to blame one international agency or another for the high tariffs—when in fact decisions on telecommunications tariffs are the preserve of national authorities. Furthermore, the space segment is the least expensive part of a satellite transmission since the heaviest costs are incurred by the use of earth stations, and the landlines to and from these stations. Also, even though the cost of using the space segment has shown a dramatic decrease in recent years, this decrease has only to some extent been passed on to customers by the telecommunications agencies which tend to regard all trans- mission systems as part of one overall network and to set tariffs so as to protect their investment—in submarine cables and other facilities also. However, appeals to international agencies such as Intelsat to lower their tariffs seemed an easier task than the more difficult and less glamorous one of achieving co-ordinated and coherent policies at home.

Recently, developments have pointed in the direction of more policy-orientated attitudes and negotiations on a more comprehensive basis. Following intensive discussions, the broadcasters have joined with the news agencies, partly under the aegis of UNESCO, for renewed discussions with such opera- tional bodies concerned as Intelsat. Thus, the news agencies stated that the flow of information between countries is impeded by unacceptably high tariffs for international news exchanges, the unavailability of services required by the press, and arbitrary regulations restricting the use of current trans- mission facilities: suggestions were made as to how to correct the situation. The regional broadcasting unions wanted to rent on a permanent basis dedicated half-transponders which would enable the unions and their members to use the satellite system on a multi-origin, multi-destination basis twenty-four hours a day. The reactions of Intelsat were favourable but a number of questions will have to be negotiated with the national telecommunications authorities.

The case of television news transmissions points up some issues which need to be considered in more general terms: how should cost-savings from a new technology be weighed against the amortization of investment in older technologies; how can we identify in time the policy aspects of questions that have traditionally been dealt with at a technical and administrative level; what social machinery do we need to co-ordinate at the national and international level issues that require an inter-departmental approach?

All these efforts seem to be for the good, but even they have to be seen in a larger perspective: the implications of all the new technologies for the production, transmission and dissemination of news. As one observer has trenchantly put it:

The dazzling array of scientific advances over the past 20 years— satellite broadcasting, light-weight electronic equipment, computer-generated graphics—have not so much changed the basic nature of television news as re-enforced it. Where television was always strong, most notably in covering major events, it has become even stronger; where television was weak, most notably in making sense of these events, the immediacy afforded by the new technology has in many cases made those weaknesses even more glaring. [Friedman, 1983.]

Information exchange: data transmission

The use of communications satellites for the transfer of information in the form of data transmission combines two of the most distinctive recent developments in communications; each represents a complex field of analysis, policy and planning. In the view of many observers, the advent of communications satellites heralded a new era in communications, characterized by an increasing rate of technological innovation and the rapid introduction of new electronic communications and information services. Similarly, the marriage of computers and telecommunications to form the new field of data communications—sometimes also referred to as tele-information systems, telematics or informatics—is regarded as a key factor in the current transformation of the information environment. Before proceeding to an account of the use of satellite services it would be useful to indicate some features of the use of telecommunications facilities for data transmission.

A total tele-information system is made up of several major sub-systems, including data transmission, data terminals and, most frequently, data processing or computer facilities. Obviously, these sub-systems must be thoroughly integrated to form a complete tele-information system but often the planning is divided into these sub-systems of which data communications is the one of immediate relevance.

The development of data communications is often seen to originate with the recognition in the late 1950s that there was a need for communicating data involving larger volumes of information at higher speeds than could be accommodated by the then current telex service. It was also realized that the increasing automation of the telephone network could considerably improve its transmission performance and that the use thereby of the telephone network would provide a better ratio of cost to performance for the transmission of data. In the international perspective, this development was conditioned by the action within the framework of the International Telecommunication Union (ITU) which agreed on a series of basic recommendations on data transmission. These standard-setting recommendations greatly assisted the growth of computer-to-computer, terminal-to-computer and terminal-to-terminal communication.

By 1970 it had become clear that data communication was growing at a rapid rate and that this growth could be expected to continue in the future. Thus, economies of scale could justify the establishment of separate communications net-works, specifically designed for data communications. Consequently, during the 1970s recommendations for public data networks were adopted.

The planning and implementation of satellites for data and information transfer can generally be seen as struc-tured around the concept of networks, conceived at several levels:

— networks, understood in the sense of organized user groups, each with specific communications requirements;
— networks signifying the telecommunications infrastruc-ture, which combines various transmission technologies into an integrated system, as in the case of the telephone network or a public data network;
— networks seen as encompassing the interlinkage of com-puters and users through telecommunications, often described by such expressions as 'computer networks'.

Many attempts have been made to define the desired characteristics of communication networks with particular reference to data transmission in terms of both providers and

users of information. In overall terms, one recent analysis
lists ten features required:

   (i) *Availability* is a matter of the physical existence of
communications systems; availability also implies the
existence of actual channel space and operational
readiness to transmit and receive messages 'at a time
and place whenever and however there is a message to
be transmitted'.

  (ii) *Reliability* concerns the provision of service on a regular, established and uninterrupted basis.

 (iii) *Integrity* refers to message clarity and accuracy: senders
want to ensure that the message sent is the message
received without error.

 (iv) *Security* is defined as control of access to messages;
preventing access by other than the intended recipient
is desired by some senders but not by all.

  (v) *Efficiency* is seen as an economic goal: the communications system should be cost-effective for the purposes
intended. An example is the case of satellites mentioned
earlier: some countries prefer what is called 'thin route'
capacity, i.e. high-powered satellites with low-powered
earth stations, to low-powered satellites used with high-
powered earth stations.

 (vi) *Transparency* means that the manner in which a message
is sent should not be obvious to the user. The user
should not bear any particular burden or inconvenience
according to whether the message is sent by satellite,
microwave, cable or optical fibre.

 (vii) *Interconnectivity* in this context implies more than
technical compatibility. It includes rules for access to,
through and across various public and private networks;
in this perspective public networks should not be
permitted to exclude any users (including private networks as users), and private networks should be capable
of interconnection unless there are compelling reasons
for maintaining privacy.

(viii) *Interactivity* implies the potential for immediate
response, as in the case of a telephone system; it is
pointed out that such interactive communication is

becoming more readily available both in the mass media and in the point-to-point system.

(ix) *Diversity* would provide for the choice of a variety of modes of transmission, e.g. telegraph, telephone, telex, facsimile, cable television, etc. User choice rather than carrier choice or regulatory rulings should be the prevailing principle. The sender and receiver should be able to use voice, video, text, freeze frame, graphic or data transmission, singly or in combination, according to the nature and function of the message.

(x) *Universality* refers to the ultimate goal of extending availability, interconnection and transparency, so that these features will be common to all of the world's systems and networks (see Branscomb, 1980).

These ten characteristics have obviously not been totally realized anywhere, and they might never even be so. Their importance lies in setting out ideal standards for user-oriented communications networks and in providing a yardstick against which to evaluate the performance of a network, as well as relevant policy and regulation.

In order to avoid confusion it should also be noted that a number of the expressions employed to describe these desired characteristics are often also used in a more technical and precise manner by telecommunications and network engineers. Here they are used in a rather general sense to indicate desirable overall characteristics of systems and networks.

Various types of computers and data networks have already been established in a number of countries. With regard to international networking, a distinction is made between three kinds of networks:

(i) Public data networks which are now being established throughout the world, generally by the telecommunications authorities. Thus, the Nordic Public Data Networks were first established at a national level by the telecommunications administrations in Denmark, Finland, Norway and Sweden. Following an expected increase in data communication requirements, regional interconnection was then established.

Special importance has been given to Euronet, established among the nine countries of the European Community, since it may set patterns for other regional public data networks. Euronet was established to meet the recognized need for a network which would enable terminals in any EEC country to have access to data bases held at various locations within the EEC countries. Phase 1 of Euronet has three objectives:

— to turn Euronet into a public operational on-line network;

— to develop a common market for scientific and technical information;

— to promote technology and methodology for improving information services.

The Direct Information Access Network (DIANE), which is the name given to the user interfaces of Euronet, had ensured access to about 140 data bases after one year of operation—1980–81. These include data bases for bibliographic and factual information retrieval on medicine, chemistry, engineering, textiles, agriculture, socio-economic data, aerospace, nuclear science and other subjects.

The network will also be allowing non-Euronet traffic, and countries outside the EEC could join.

(ii) Closed user group networks which comprise different types of users. These networks include the Global Telecommunications System operated by the World Meteorological Organization (WMO) and the network of exchange of criminal identification information operated by Interpol. Two other widely-known networks are SITA (Société Internationale pour la Télécommunication Aéronautique) which has some 200 airline company members and operates a network of airline reservation services in some 118 countries, and SWIFT (Society for Worldwide Interbank Financial Telecommunications) which today comprises some 1,000 member banks.

(iii) Commercial networks; the commercial network providers usually offer time-sharing services on their own computers and communications services for clients of other computer centres. Thus, organizations located long distances from computer centres may dial processing and storage capabilities or interrogate different data bases which are offered on these networks. The earliest computer service companies to offer international services were based in the United States (Mark III, TYMNET, CYBERNET, etc.) but more recently international information retrieval and data processing services have been established in Canada, France and other European countries. (See UNESCO, 1982.)

As already pointed out, satellites are used for all kinds of telecommunications services, among them information transfer, as understood in this section. Seen from the point of view of the satellite system operator, information and data transfer represents one service, one category of customer among all those serviced by his system. Generally, satellite facilities form part of the telecommunications networks established by the telecommunications authorities, common carriers or similar telecommunications operators. Thus, messages passed between two correspondents on opposite sides of the North Atlantic are routed via the terrestrial network to the sending, and from the receiving, earth stations, or via undersea cables. Data and other information is routinely transmitted over available channels, with the customer being generally unaware of, and probably uninterested in, which facilities are being used. This situation is, however, changing, through the use of satellite systems which provide services directly to the customer, whether as a direct broadcast satellite service, or as new two-way communication services.

In the following pages a number of different cases in which communications satellite facilities are used for the purpose of information and data transfer will be presented.

Palapa

Indonesia was the first developing country to establish an operational domestic satellite system. The Palapa system was introduced in 1976 in order to meet the combined requirements of Perumtel, the Indonesian telecommunications company, and other users. The satellite option is particularly suited to the geographical conditions of Indonesia, in linking the 13,000 islands which make up the country.

The Palapa system provides a complete range of telecommunications services including telegraphy, telephony, telex, data transmission, television and radio transmissions, which are available to both public and private users. A particularly compelling reason for the establishment of the system was the possibility of providing reliable communications services for industry and business: the national oil company, banks, insurance companies and other enterprises needed to communicate with their branch offices, customers, etc., dispersed throughout the country.

Based on preliminary traffic studies, fifty ground stations were planned at the introduction of the system. There were four types of station in use:

— one master control station;
— eighteen main traffic stations;
— fifteen light traffic stations with television facilities and six without television facilities;
— ten industrial stations.

While data can be transmitted over all these facilities, the industrial stations are specifically intended for data communications: they are designed to be compact and inexpensive (thirteen-foot parabolic antennas and transistorized receivers).

As the Palapa A satellites approached the end of their design life, the Indonesian telecommunications administration implemented a plan to launch a new series of satellites, which began in March 1983. The new generation of satellites will double the transponder capacity and will continue the expansion of telecommunications services. The number of ground stations will be considerably increased to a total of 225 stations of various types; seventy-five of the envisaged one hundred TV receive-only stations are intended to provide rural service.

In the case of the Palapa systems, data transmissions and other forms of information transfer are one class of service within a system which is integrated with the general telecommunications network and is used for the normal mix of telecommunications services.

PEACESAT

A very different approach is represented by the PEACESAT project (Pan-Pacific Education and Communication Experiments by Satellite). The PEACESAT project was initiated by the University of Hawaii and started experimental operations in 1970. It represents one of the first experiments in the use of low-cost earth terminals for two-way communication by voice, teletype, facsimile, computerized data transmission and still pictures. The National Aeronautics and Space Administration (NASA) provided voice grade circuits free of charge over the experimental satellite ATS-1.

The project was designed to interconnect educational, research and other similar institutions in the Pacific Basin. The terminals, which were installed in some sixteen locations, were of several kinds to meet specific social requirements:

— a portable terminal—light in weight—sends and receives voice communication, for research teams, regional inspectors, patrol officers and medical personnel;
— a standard terminal, permanently located, sends and receives voice, teletype, facsimile, intended for small institutions, medical centres, schools and community centres;
— a full utility terminal interconnected with libraries, classrooms, medical centres, etc.;
— a receive-only terminal—small in size—intended for students, news and remote personnel.

Costs ranged from $7,000 for the larger terminals to $100 for the receive-only terminals.

The specifications of the experimental system included the following:

— terminals mainly designed for both reception and transmission of messages;
— local ownership and operation of ground facilities;
— low-cost and easily-moved facilities;
— multi-media centres with direct access to other centres;
— equipment that could be operated by local personnel with only limited special training;
— a global service area, covering a third of the earth's

surface, providing sufficient traffic to justify service in 'low density' areas.

The objectives of the project were to improve communications for education, health and community services in the Pacific. The system has been used for instruction on a regular basis, health and medical consultation and exchange, conferences, seminars and other exchanges among specialists in agriculture, health, education and culture, news and information exchanges.

Of particular interest are the opportunities offered by a system such as PEACESAT to establish networks of varying sizes—networks being understood as organized groups of system users—each with specific communications requirements. Thus, on the basis of the experience gained in exchange between libraries, several types of library networks were considered. Since library traffic is not so great as to sustain a separate, dedicated long-distance transmission system, the sharing of facilities with other users becomes important.

PEACESAT is an interesting example of providing satellite facilities in keeping with the idea that small is beautiful. The growing interest in the Pacific Basin area as an increasingly important geo-political region has also led to proposals for the introduction of communications satellite systems in line with the notion that 'large is beautiful'. Intelsat, American companies and Japanese agencies are all proposing various versions of a Pacific regional satellite system—without, however, any visible investigation as to what the Pacific Islanders themselves might need—or want.

PADIS

In the case of Palapa and PEACESAT, existing satellite facilities intended for a variety of purposes have also been used for data transfer. In other cases, the planning of data transmission systems specifically includes the use of satellite facilities.

An important example at the regional level is the Pan–African Documentation and Information System (PADIS) which was formally created in 1980, within the framework of the United Nations Economic Commission for Africa (ECA).

The main operational objectives of PADIS are to provide ready access to development information for policy-makers, technicians, planners, financiers and others engaged in the economic and social development of African states. This will be achieved by:

— identifying and collecting African information resources and creating an efficient system for the utilization of the information by ECA and the member-states;
— promoting information exchange;
— ensuring access to both published and unpublished documents produced in Africa on all questions relating to economic, social, scientific and technological development;
— providing assistance to member-states of ECA to strengthen their national information infrastructures in order to enable them to participate fully in PADIS;
— establishing links (systems interconnection) with international information networks, data bases, information and documentation units situated outside Africa as additional sources of development information;
— providing a Pan-African information system utilizing the most recent technology of data transmission, including communications satellites.

The PADIS telecommunications network will include the following basic programme:

— installation of a remote interrogation terminal at the CCO (Central Co-ordinating Office) interconnected with the ESRIN data bank established by ESA (European Space Agency) at Frascati, Rome, via direct point-to-point telecommunications channels, through satellite facilities;
— design and implementation of a Pan-African data transmission network by using communications satellites, establishing sub-regional network nodes and covering all the member-states. The installation of such advanced technical facilities at the regional, sub-regional and national levels could also serve other applications for which transfer of knowledge and dissemination of data are important.

The project will be implemented in three phases during the period 1980–88, and will require an overall budget of $145 million. The data traffic has been estimated at a total of 150,000 connect hours per year for 200 access points in about fifty African countries, representing a system connect-time of 3.75 hours/day per access point. This would correspond to some 10 million printed lines per day. According to these estimates of the data traffic, the communications capacity would in itself justify the setting-up of a satellite capability. However, it is foreseen that this satellite system should also serve other needs which have been identified by ECA, in particular a pilot project for the distribution of educational and television programmes across the African continent.

A total of 102 ground stations for connection with the satellite has been envisaged. These ground stations would be small (1.5 or 3 metre parabolic antennas). The entire cost for the installation and operation of these ground stations has been estimated at about $33.35 million. This sum includes the cost of two operators per ground station during two years as well as the training of these operators, project management costs for the installation of the stations and local staff training (see United Nations Economic Commission for Africa, 1980).

DEVNET

The implementation of PADIS will depend on the use of satellite facilities to link the African regional and sub-regional centres and to connect the African network with data bases and networks in other continents. Similarly, the implementation of the UNDP-proposed Development Information Network (DEVNET) is predicated on the use of satellite facilities. This is envisaged as a multi-regional network which will primarily consist of teleprinters linked by satellite channels capable of simultaneous two-way transmission to and from a computer-operated communications centre, supplemented by landlines in countries with satellite ground stations. In the planning, priority will have to be given, for technical as well as for practical reasons, to countries with access to satellite communications channels.

The objective of the network is to provide full horizontal information flow between developing countries to promote and support co-operation for development. This requires a computer-operated network which provides a high processing and switching capacity, with considerable built-in flexibility at reasonable cost. The network should use a pre-coded distribution system for 'addressing' information by computer to particular countries, regions or categories of users, so as to handle the multi-directional flow automatically and yet permit users themselves to pre-select a manageable quantity of information from the service.

The computer-operated section of the network will function at two levels:

— regional: a star network will provide two-way communication between countries in a given region or sub-region through a central processing and switching facility located in the regional (or sub-regional) linking centre;
— multi-regional: a 'distributed' network will provide complete multi-directional communication between regional centres, and hence from any country linked to a regional centre to all others similarly linked.

At the national level users can be linked to the national dissemination and retransmission centre by teleprinter line or other delivery mechanisms. The network would have a decentralized structure at the regional and sub-regional levels. Progressing step by step, six regional or sub-regional linking centres would be put into operation in the following areas, defined according to cultural rather than purely geophysical criteria: (i) Latin America and the Caribbean, (ii) the Arab world, (iii) South Central Asia, (iv) South-East Asia and the Pacific, (v) East and Southern Africa, and (vi) French-speaking Africa. It is foreseen that at the end of the three-year project, the network should consist of the following:

(a) six regional linking centres, which would be the structural pillars of the system;
(b) national dissemination and retransmission centres, connected by satellite to the regional or sub-regional centres, which would feed and receive material from the national

level into the network and direct the incoming flow of
information to users within a country;

(c) individual users, linked directly to the national centre,
for two-way flow of information;

(d) a network co-ordinating office.

With regard to cost, the projected telecommunications
costs are of immediate interest in this context. For a three-
year period the telecommunications costs for each regional
(or sub-regional) centre have been calculated to be no
more than US$65,880 and for each national centre to be
$6,480 (corresponding to $2,160 per year for one satellite
link-up. (See United Nations Development Programme, 1980.)

SBS

In terms of satellite systems providing data transmission
services direct to their customers without passing through a
general telecommunications network, the American Satellite
Business System (SBS) is an interesting case. SBS is a joint
business venture of COMSAT General, IBM and Aetna Life
Assurance, to provide a commercial service to large numbers
of customers who will own and operate their own earth
stations. It is in fact a new United States common carrier.
The intended service is to be entirely digital and will offer
customers private networks, fully switchable for integrated
high-speed voice, data and image communications.

Integration tests began in 1980 while an operational
service started in January 1981. The system consists of two
satellites launched in 1980 providing ten communications
channels. There is an option for six additional spacecraft.
The system is aimed primarily at large business customers
requiring intra-company communications. It is designed to
operate with relatively small earth stations (5.5 metre or
7.7 metre diameter antennas) situated on user premises.
The demand for service has, however, not met expectations
and a considerable part of the capacity is used to serve
smaller users via 'city centre' earth stations and 'on premises'
stations (service points). Services offered include voice
channels, teleconferencing facilities, high-speed facsimile
(seventy pages per minute), and low-error rate data transmission.

It is expected that satellite systems offering these kinds of direct services to the end-user will become more common in the future. One of the French national satellites will provide similar services and there are plans for Intelsat and European satellites also providing this kind of specialized service.

Education by satellite

It is only fair to start with the first, basic question: what do satellites have to do with education at all? The answer is complex and has to begin with present educational problems: the crisis *of* education and the crisis *in* education. All countries and, in particular, the developing countries, are in the throes of an educational crisis. There are many and varied reasons for this crisis; they can be seen as both external and internal to the educational system. External to education is, first, the growth in demand for learning, resulting both from the increase in population and the shift in demographic composition towards a predominance of young people in many developing countries. At the same time, there has been an increase in expectations, expressed in such commonly-accepted goals as: the eradication of illiteracy, primary education for all, and the democratization of higher education. The population explosion is matched by a knowledge explosion and the exponential growth of available information. Thus, it is increasingly recognized that conventional systems of learning can no longer absorb the range of knowledge generated, disseminate it in the usual time-span or respond adequately to demands for equitable and widespread access to knowledge and information or to the learning needs caused by rapid out-dating of knowledge. New modes of education, learning and sharing of knowledge are therefore needed, at all levels of society, using a wide gamut of services and techniques.

Concurrently, the internal crisis in education has been expressed in vastly changed concepts of education. The critique of traditional attitudes has, in an extreme form, been formulated in Ivan Illich's plan for the 'deschooling of society'. More acceptable has been the idea of education as a permanent, life-long activity, as envisaged in a well-known UNESCO report with the revealing title 'Learning to Be'.

More recently, a report to the Club of Rome entitled 'No Limits to Learning' promotes the idea of innovative and integrative learning which requires the addition of two major features. The first feature, anticipatory learning, is seen as promoting solidarity in time, by using anticipation as an ability to face new, often unprecedented situations and to create new alternatives where none existed before. The second feature, participatory learning, creates solidarity in space: the aim is to foster participation in the learning and information-sharing processes at all social levels and at all ages. A further extension of learning is represented by the concept 'global learning', as envisaged by The United Nations University. It expresses the need for developing not only individual learning capacity but also the learning capacities of institutions and even societies and of improving social preparedness and the ability to cope with growing complexity at the global level. We need therefore to learn how to generate, evaluate, select and share vast amounts of new and relevant information.

It is against this background that new educational technology and the use of new technologies for education have been hailed as valuable advances. Great hopes were placed in the use of educational cinema, educational radio and educational television. In this emphasis on the distribution of education, the satellite option was soon given a special importance. In fact, in the late 1960s and early 1970s there were high hopes that satellite broadcasting would provide a solution to some of the most pressing educational and development problems of Third World countries. A satellite does enable a broadcasting and distribution system to be set up much more quickly than would be possible by conventional methods. This was seen as having particular relevance to developing countries which do not have an extensive ground infrastructure, especially in remote rural areas. Also, given the educational problems in developing countries, time is a crucial element. Satellite-based systems can not only provide services more quickly and thereby attack the quantitative aspect of the problem but they can also bring about a change in the quality of education by making the best teaching available to all. More innovative use of satellite capabilities can also bring about altogether new modes of education, especially for specialized instruction.

Initially, television signals beamed via satellites could only be received by large, complex and therefore expensive earth stations from which the signals could be sent to conventional television transmitters for re-broadcast. Since reception of such signals does not require any additional technical equipment beyond the normal television set, this approach—the distribution mode—is still useful and economic in areas with high receiver density, e.g. urban areas. Satellite distribution of television programmes in this mode has been used extensively for national distribution of programmes in Canada, the United States, the USSR and Indonesia. However, as noted earlier, rapid advances in satellite technology soon made possible what has been called a 'technology inversion' in which the spacecraft became larger, more sophisticated and complex, thereby enabling the reception equipment on the ground to become simpler, smaller and cheaper so as to enable direct satellite broadcasting without passing through the intermediary earth stations, landlines and television transmitters. The first major example of this technology was the ATS-6 spacecraft in the United States. Using this satellite it became possible to broadcast directly for reception by cheap and fairly simple ground receivers. In both the United States and India the television signals from ATS-6 were received on a 3 metre diameter parabolic antenna. More advanced satellites have been used with ground receivers using anntennas of about 1.2 metres diameter. And already satellite systems are being developed which will use receiver antennas of 0.9 metres and even smaller.

Thus, among the technical possibilities now offered by satellites are: radio and television distribution to broadcasting stations; direct television broadcasting to augmented sets; direct television broadcasting with two-way audio, two-way video and audio; slow-scan television with or without two-way audio; television with teletext; facsimile; access to computers for computer-aided instruction or for other purposes.

While operational satellite systems that can offer all or most of these services are already in place in the United States, the USSR, Canada and Indonesia, none is used extensively, or even primarily for education, but rather for revenue-earning telecommunications/data traffic and for relay or exchange of normal television programmes.

New communications satellite systems are planned or proposed for India, Europe—France and Germany and the Nordic countries—Saudi Arabia, Latin America, the Arab countries, Japan, China and others. Most of these satellites will have television broadcast capability but the main thrust is likely to be in the telecommunications area. The extent to which these systems will be used for educational and instructional purposes is uncertain. Even so, the fact that so many systems are under preparation is an opportunity for policy-makers, educationists and others to ensure in the planning stage itself that an educational element is built into the patterns of use. For other countries this may be an opportunity to examine whether an educational satellite system would be useful and economic.

There are a wide variety of reasons for the slow progress of satellites for educational purposes. One obvious problem is the high investment required not only for the satellite and its launch but also on the ground. Despite the growing recognition of the importance of education, many governments still tend to look at education as an overhead rather than as a high-priority investment—probably because the returns are not immediately and concretely visible in quantitative terms. One solution to this problem is the multi-purpose satellite system in which the instructional use can, as it were, ride piggy-back on the more economically attractive services like telecommunications. This is the case with the Indian Insat and the proposed Arabsat. In fact, most countries now planning their satellite systems seem to favour this approach. A second solution would be for the investment in the spacecraft and launch to be shared by a number of countries. This is an attractive idea for smaller countries and a 'regional' approach is being considered by various sets of countries, including the Arab countries, the Nordic countries and some South American countries.

Another problem is the lack of familiarity with satellite technology on the part of the educational planners and educationalists in most parts of the world. In contrast, the telecommunications personnel have always been technology-orientated which is one reason for the quick entry of satellites in their field. A third and major problem is related to the issue of the practical efficacy of educational and instructional television. This is still an unsolved problem. The educational uses of new media, of cinema, radio and television have sometimes been a resounding success, but have equally often disappointed the great hopes once placed in them. There is still a debate about the best manner in which these new instruments can and should be used, about the educational purposes they can serve. While there is as yet no one answer to these questions, it does seem reasonably clear that it is not enough just to graft a new medium on to existing educational structures and methods. Only when education itself has been re-thought in content and methodology, when educational systems have been opened up to reach new groups in new ways—as expressed in the concept and practice of open universities and distance learning systems—do we seem to get closer to realizing the educational potentials of new media.

Another basic problem with regard to the setting up of instructional satellite systems is one of structure. It is well known that the integration of a major new element in to any system requires a change in the mode of functioning and often in the structure of the system itself. Satellite broadcasting can be added to the existing educational systems only when certain operational and structural changes are made. The type and extent of change will obviously vary with circumstances. However, the stability or—seen from another angle—the inertia or resistance to change of the educational systems is a fact of life in all countries.

At the same time, the magnitude of the educational problems, especially in the developing countries, is so great that conventional techniques and approaches are unlikely to succeed. This has been brought out in studies on the feasibility of using satellites for development and education, undertaken in India, Indonesia, Pakistan, the Arab countries, Brazil and the Spanish-speaking countries of South America. In the case of India these studies were followed up by practical experiments. In other cases they were supplemented with thorough exploration of the future of the entire education system, such as the UNESCO/UNDP study on tele-education in South America, the SERLA project. Even in developed countries, satellite-based systems might be required. The Educational Communications Authority of Ontario in Canada has stated that it will use satellites to distribute its programmes.

SITE

Thus, the desirability and feasibility of educational satellite systems have not remained merely theoretical studies; there has been demonstration of them in practice. The most extensive experiment so far has been the Indian Satellite Instructional Television Experiment (SITE) carried out for a year in 1975/76 using the AT-6 satellite. The same satellite was earlier used in the United States for experiments in the fields of health and educational delivery services.

The lessons drawn from these experiments are particularly important with regard to the planning of the system and its operation. An instructional satellite system consists of a large number of component parts working in unison to provide a defined service. The most important point to note is that the various sub-systems must be planned around the projected usage. It would be counterproductive to establish a system first and then to plan its usage.

Expressed in basic terms, the operational goal of an instructional satellite system is to provide a broadcasting service to specific target audiences. Thus, SITE had as its goal the provision of a television broadcasting service to 2,400 villages in six clusters spread over India. The target villages were selected for electrification—even though some 180 villages in Orissa used battery-operated receiving sets— and accessible for 80 per cent of the time by road from a nearby town so as to enable maintenance to be carried out. The target audiences were schoolchildren in the five to twelve-year age group in the morning and adults in the evening. A total of four hours of programming was available per day.

According to the Indian experience, the five main components of an instructional system comprise: programme transmission; programme reception; feed-back and evaluation; organization and management. The programme production component consists of studios and the associated production staff. This would include fixed and mobile studios. The programme transmission component includes satellite earth stations and their feeder links, the satellite itself, and in certain cases terrestrial transmitters or cable systems. The programme reception component consists of the broadcast receivers and the programme utilization elements. The receivers can be of two types: conventional television sets or specifically augmented satellite receivers. The feedback and evaluation component consists basically of social research staff. The organization and management segment is the nerve centre which organizes, directs and co-ordinates the operation of the other four components.

In the context of SITE the measures taken for the receivers are of special interest. The SITE direct receivers consisted of a specially-designed antenna made of chicken wire and an electronic system called the 'front converter' to receive the signals from the satellite and generate audio and video signals for feeding into the television receiver. The receiver was a commercially available unit modified to interface with the converter and to tolerate rough field usage. The major modifications were an enlarged cabinet to protect the picture tube, particularly during rural transportation, a power supply which could withstand the voltage fluctuations expected in rural areas, and a patching system whereby the set could be converted from direct satellite reception mode to conventional VHF mode. The antenna and the converter were new designs specially made to reduce costs and ease maintenance in the field. The antenna was in kit form, which could easily be erected in the field by two people using simple off-the-shelf tools.

The site selection for the community receivers was perhaps the most time-consuming process. It started by first selecting the target villages where community sets were to be located. This was followed by a visit to each of the villages to determine the accessibility of the village, a suitable building for the community receiver and nearest electricity point (except of course in the case of battery-operated sets). Accessibility had direct bearing on maintenance. Though in principle all villages should be served, inaccessible villages would severely strain the maintenance system. In community viewing situations in developing countries the suitability of a building to house the community set is decided by three criteria: it should be a public place capable of serving two to three hundred persons; it should provide security to the installation and it should have near it a suitable site for erecting the antenna.

While there are a number of constraints and regulations under which a satellite system has to operate, there are, even with these constraints, a wide range of choices to be made. It is these choices that result in the tailoring of the system to the specific needs and resources of a country. Thus, the system design of an instructional satellite system is a complex process comprising a large number of variables and trade-offs. The parameters involved include those relating to the satellite, to the ground hardware, to the programme design and production system, to the maintenance system and to the overall organizational structure. While some of the trade-offs involve straightforward technical and economic choices, others are qualitative and can be resolved only on the basis of 'subjective' preferences. System design has to take into account political, social and cultural factors, in addition to the more obvious technical and economic factors. The choices concern selection of areas to be served; examination of whether a country would like to be part of a regional satellite system or would prefer to have a dedicated satellite for its own use; choice of medium—which depends on such factors as size, level and type of audience, messages to be conveyed and objectives; mode of reception (community or individual); existing infrastructure for television, radio and telephony in the area; investment constraints; physical constraints in terms of power, viewing and listening space, and maintenance; definition of message content; number of channels; intensity of coverage in terms of the proposed number of receivers within the given service area, etc.

As in any venture, the most careful planning can come to nought without proper management and organization. In an instructional satellite system the very nature of the system and the sophisticated technology involved require exceptionally good management, combined with proper organization. These must therefore be treated as important and critical elements in the overall system design.

Social and cultural factors are primary determinants in the design of the management systems and organizational structures. Therefore, just as the hardware and programmes should be custom-tailored for the needs of each country so the management and organization structure have to be designed specifically for each situation.

In an instructional satellite system, there are generally a number of agencies involved, which demands a high degree of co-ordination:

— the agency dealing with space: in some countries a separate agency while in other countries it is the department of communications or the telecommunications administration which is responsible for communications satellites;
— the education sector, including all sectors requiring delivery of messages, such as agriculture, health, family planning, etc.;
— the broadcasting agency/agencies;
— the telecommunications authorities and/or agencies.

An advanced satellite broadcasting system demands an advanced software system. The primary target audiences of the instructional programmes in developing countries will be the socio-economic and culturally disfavoured populations living in rural and remote areas which are usually not well served, if at all, by agencies for education and development.

The wide coverage of the satellite leads to physical separation of the designers and the receivers of the programmes. But perhaps a more serious problem is the perception-gap in respect of the real needs, aspirations, background and beliefs of the rural audience. The production team is more often than not likely to have special ideas and orientations about the role of the medium itself. In the case of urban programme designers producing programmes for rural audiences, development itself has different connotations according to the two groups.

Though handicapped by their background and training,

the producers, researchers and experts who constitute the information source must be willing to learn from and to develop an empathy with the rural audiences. It is in this context that the idea of 'formative research' has evolved. The aim of this research which should be carried out by appropriate social scientists (sociologists, anthropologists, psychologists, communications experts) in the area to be served is to sensitize the producers and to help them develop empathy with their audiences. Production should therefore be carried out in a team mode.

The programme designers will have to acquire background information about the audience which should include their psychological make-up, socio-economic beliefs, superstitions, customs and practices. For this purpose detailed 'audience profiles' will have to be prepared after extensive fieldwork. The audience profile would contain information on climate, population, social structure, current informational and educational levels, economic activity, agricultural practices, animal husbandry, health and family welfare, cottage and small-scale industry, planning and development, social life, etc. The next step would be to establish the perceived and real needs of different socio-economic categories of people in the instructional areas. There should be special emphasis on the relevance of the educational content, its utility, practicability, timeliness and location specificity.

If the full potential of an educational satellite system is to be realized the software system must be put in the context of the development process itself. In order to gain credibility and to achieve the stated objectives the medium should be handled in such a way that the audiences identify the medium as their own, and one to which they can have access to ventilate their grievances, to close the information-gap and to establish two-way communication between them and the decision-makers. Their involvement and participation in the very process of programme design leads to improvement in their self-confidence. To this should be added the need for different development agencies (in agriculture, health, etc.) to use instructional broadcasting as an extension tool. To provide for audience participation and use by development agencies, a conscious effort will have to be made to bring down the cost of equipment and production of programmes, especially in the case of television. It is also necessary to demystify the medium for potential users. In this context the choice of appropriate technology becomes crucial. An expensive, large studio with sophisticated and immobile equipment will necessarily lead to a high degree of studio-based programmes. On the other hand, a decentralized system with a number of small studios with portable, light-weight equipment will result in a substantial number of field-based and participatory programmes.

As could be expected in the case of such an innovative project, opinions about SITE have been extremely varied. Some have criticized the use of a centralizing, high technology for development purposes in rural areas; others have levelled detailed criticism against specific programmes. It would seem though that some of the most positive results were unexpected. Thus, an Indian expert has stated that:

for the first time the entire cross-section of the village community had access to the message in reality. For once a communication situation was created where the less privileged did have equal access to information with the elite . . . Caste distinctions tended to break down as people were forced to sit closer together to view programmes on the 23-inch screen . . . a spin-off from a community education input that is equally accessible to rich and poor, literate and illiterate. [Krishnamoorthy, 1977, p. 72.]

And an international evaluation team from the Commonwealth Broadcasting Association had this to say:

Perhaps the most striking outcome of SITE so far has been the new-found confidence that has emerged amongst all those who have been involved in the experiment. The doubts and anguish that overwhelmed many of them in the period leading up to SITE have evaporated in a euphoria of delight at the measure of success that has been achieved so far: systems have worked, programmes have been produced on schedule and problems have been dealt with. [Quoted in Krishnamoorthy, 1977, p. 71.]

And a recent United Nations study reaches the following conclusions:

Progress in setting up specially dedicated systems for education has been somewhat slow. This is natural. First, because even the use of conventional telecommunication and broadcasting for education is not as widespread as had been hoped a few decades ago. Secondly, the education and telecommunication organizations in most countries work under their historical constraints and limitations while the use of space technology of education is somewhat difficult without new organizational structures. Thirdly, the increasing range and reach of satellite signals present genuine problems and challenges for devising educational curricula to suit the needs of different cultural and language groups. Fourthly, new mechanisms have to be devised for debiting relatively large initial costs of space systems which, instead of being used as an additional investment in education, can also be used for revenue-earning telecommunication services. These are genuine problems, but all can be solved. [UNISPACE 82, Document 101/BP/6, p. 31.]

And India for one seems intent on acting accordingly. Through Insat, the multi-purpose national satellite launched in 1982, the Indian public television corporation started to beam programmes to low-cost transmitters and receivers in remote villages all over the country. By the end of 1982, there were 6,000 community receivers, costing the government no more than some $1,200 each. As in SITE, programmes in the morning are intended for schoolchildren and in the evening for adults. Another country which seems ready to take a similar path is China. Some 10 million people leave senior high school each year and the universities by themselves cannot meet the demand for higher education. So the government is planning a television satellite to be operational within the next ten years—this lead time is required to produce in the country the sheer quantity of ground equipment required.

Obviously, satellites by themselves cannot solve educational problems. They can, however, provide an important tool for specific learning and information purposes. To take advantage of the satellite potential it is necessary to include the educational aspects and applications at the early conceptual and planning stage. Only then can the revolutionary development of space systems 'ultimately find an important place in the instructional activity of this planet' (idem, p. 32).

6   The great satellite game: round 2

The second round of the great satellite game can be characterized by the word 'more': more satellite systems, more
actors, more purposes, more problems. In the first round
there were two main actors: the United States and the
USSR. Little is known of the internal struggles in the USSR
about the directions that the space and satellite programmes
should take. The United States scene is more open so that it
was possible to identify the interest groupings involved and
to follow all the sub-plots, many of which have had immediate
international repercussions. Other countries seemed merely
to react. In effect, most of their initial actions were no more
than reactions to NASA's proposals for co-operation and to
the American push towards the establishment of Intelsat.

   The second round is very different. More and more actors
are appearing on the international scene, with their own
policies, purposes and programmes. An early entrant is
Canada, which managed to design a relatively coherent
national policy and establish both the social instrument for
and the first national satellite system using geo-synchronous
satellites, the 'Anik' system. Japan proceeded, apparently
very calmly, with an ambitious programme of technical
development and the adoption of a long-term programme for
national satellite systems of increasing sophistication. The
Europeans had a difficult time getting their space programme
going and for a long time the results were confusion, with the
only practical result being the Franco–German experimental
satellite Symphonie. Of the developing countries, the first
arrivals on the satellite stage were India, with an interesting
programme centred on development of modest national
launching facilities, manufacturing of scientific and earth
resources satellites and the SITE experiment which led to
a decision to establish a national communications satellite

programme; and Indonesia which, with the establishment of the Palapa system, became the first developing country to operate a domestic satellite system. Other developing countries have taken decisions or are in the planning stage for the establishment of systems. A great number of countries have also leased facilities from Intelsat for domestic use.

Summary of present and planned communications satellite systems

Although the situation fluctuates, an overview of the present operational and planned communications satellite systems in the world would look something as follows:

1. *International communication satellite systems*

There are at present three international communication satellite systems:

(a) Intelsat, which now provides international public telecommunications services in over 145 countries and territories, and domestic lease facilities to some 35 countries.
(b) Intersputnik, currently providing various telecommunications and broadcast transmission services between the socialist countries of Eastern Europe and to Mongolia, Algeria and Cuba.
(c) Inmarsat, which has been established to provide maritime communication services to its members, now numbering about 37 in all parts of the world.

2. *Regional communication satellite systems*

(a) Afrosat. African Satellite System which is being studied by the 38 members of Pan-African Telecommunication Union (PATU) with UNDP financial support. This is one of the priority projects under study as part of the African Decade of Communications and Transport Development (1980–90).
(b) Andean Satellite System. The regional telecommunications organization of the Andean countries, grouping Bolivia, Chile, Colombia, Ecuador, Peru and Venezuela,

has for some years been studying various means of meeting the telecommunications requirements of the sub-region through satellite facilities. However, no firm decision seems to have been taken.

(c) Arabsat. Arab Satellite System. The Arab countries have decided to launch a regional system and have established the Arab Satellite Organization with headquarters in Riyadh. The system, which will cover some 20 nations, is expected to be operational in 1985.

(d) The European Communications Satellite (ECS) is a project of the European Space Agency and Eutelsat, the joint satellite organization set up by the European telecommunications administrations. The system is intended to supplement terrestrial facilities for long-haul communications, primarily for telephone and data traffic and television transmission. This has been preceded by the experimental satellite OTS and will, in principle, be followed by the L-Sat project where L stands for large.

(e) Nordsat. A Nordic satellite system has been discussed since 1975 between the five Nordic countries (Denmark, Finland, Iceland, Norway and Sweden). The intention was to provide facilities for the broadcasting of the television programmes of each country to all the others. Difficulties in reaching agreement have moved Sweden to decide on a national satellite system with some participation from neighbouring countries.

(f) An ASEAN regional satellite service has been discussed among the ASEAN countries of Indonesia, Malaysia, the Philippines, Singapore and Thailand. The plan is for use of the Indonesian Palapa satellite systems for a limited international regional service connecting rural communities.

3. Domestic satellite communication systems

These systems have been grouped in two categories: the first group comprises those countries that have established or are planning to establish domestic systems of their own. The second group includes those countries where a domestic service is provided through the lease of facilities on an existing system, primarily the Intelsat system.

Group 1

(a) Australia is now providing a service to some 60 earth stations through two transponders leased from Intelsat. However, the intention is to establish a national system to provide telecommunications and television services, primarily to the interior and sparsely-populated areas (1985).

(b) Brazil has adopted an approach similar to that of Australia. At present, facilities are leased from Intelsat but a decision has been made for the establishment of a domestic satellite system for telecommunications and television services (1985).

(c) In 1973 Canada established the first geo-stationary satellite system for national telecommunications purposes and for television distribution, under the management of Telesat Canada. The latter versions of the Anik satellites provide direct broadcast capability.

(d) China has leased facilities from Intelsat and has been planning a domestic system for the mid-1980s, particularly for educational broadcasting. Present status unknown.

(e) Colombia has leased transponder capacity from Intelsat since 1978 to provide services from the mainland to San Andres island. Plans have been announced for a domestic system (Satcol) intended mainly for rural telecommunications; however, the new government which came to power in 1983 seems to be reconsidering the project.

(f) France will have two domestic systems: Telecom 1, a public telecommunications and business communications satellite system established in 1983, and TDF Broadcast Television Satellite system for 1984/85. France has also leased circuits from Intelsat to provide telecommunications and television services to French overseas departments.

(g) The Federal Republic of Germany will set up the other half of the jointly developed Franco–German direct broadcast satellite system, for domestic television services.

(h) India has for some years leased facilities from Intelsat but in 1983 she launched the first satellite in a domestic satellite system, Insat, for community television, tele-communications and meteorological services.

(i) Indonesia was the first developing country to launch its own domestic system, Palapa, in 1976, for telephone, data and television services to a constantly increasing number of earth stations, spread all over the country.

(j) Italy has seemed to be moving towards the establishment of domestic fixed-satellite and broadcasting-satellite systems; it is not quite clear at the present moment, however, just how firm these plans are.

(k) Japan followed an experimental, Japanese-produced telecommunications satellite system with an operational system of the same kind. Similarly, an experimental direct broadcast satellite will be followed by an operational system in 1985.

(l) Luxembourg has announced plans for a 'domestic' satellite system which would in fact cover a large part of Central Europe. At present it is uncertain whether Luxembourg will establish a system of its own or will lease facilities from the French broadcast satellite system.

(m) In Mexico plans for a domestic telecommunications and television distribution system have been announced but their present status is unknown.

(n) Saudi Arabia has announced the intention to launch a national broadcasting satellite system; however, it is not clear whether preference will finally be given to the lease of facilities on Arabsat.

(o) Sweden has recently decided to launch a national satellite system (Tele-X) for telecommunications, data traffic and television services; this system may also be used by some of its neighbouring countries.

(p) In Switzerland a commercial broadcasting satellite system capable of covering large parts of neighbouring countries has been proposed, as well as a public-service oriented system. No firm decision seems yet to have been made.

(q) The United Kingdom has decided to launch a direct broadcast satellite system in the second half of the 1980s.

(r) As could be expected, the United States has launched a number of domestic satellite systems of various kinds for telecommunications, data transmission, television and cable distribution; plans for a number of direct broadcast systems have been announced.

(s) In 1966 the USSR launched the world's first national satellite system under the name Molnya, for telecommunications services and television distribution. This system has been followed by several others, including geo-stationary satellite systems for telecommunications, television, maritime and aeronautical communication.

Group 2

Among the countries that now lease facilities, mainly on Intelsat, for domestic services, or are in an advanced planning stage, are Algeria (since 1975), Angola, Argentina, Bangladesh, Bolivia, Cameroon, Chile, Denmark (for services to Greenland), Ecuador, Libya, Malaysia, Mali, Mauritania, Mexico, Morocco, Niger, Nigeria (since 1976), Norway (since 1975), Oman, Pakistan, Papua New Guinea, Peru, Portugal, Saudi Arabia, Spain (for service to Canary Islands), Sudan, Thailand, Venezuela and Zaïre.

It should also be mentioned that there are a number of experimental systems launched by the United States, Canada, ESA, France/Germany, Japan and Italy. Moreover, the United States and the USSR have launched a large number of satellite systems for military communications purposes.

Satellite policy

Satellites represent a highly visible technology—in terms of controversies surrounding organization, control and investment. They are perceived as having a great impact on communication patterns, they serve as a catalyst for new technologies and have important social, political and cultural implications. They also had a catalytic effect on a number of issues that might otherwise not have been so clearly perceived or so intensely discussed, whether at the national or international level. It is therefore revealing to analyse how different countries have approached the problems of dealing with this new technology, how they have—or have not—

organized the required co-ordination and have been able —or unable—to manage the new issues raised by the advent of satellite communications.

From one perspective satellite communications are of major significance in the area of policy. Even more than other new communications systems, they demand a new approach leading to a more comprehensive framework. It is not simply a question of how to cope with the introduction of a major new technology. It involves the capacity of national societies and the international community to deal with new areas of human activity that are multi-disciplinary and inter-organizational. Also, we must accept an inter-dependence both functional and geographical, for which our traditional categories, whether conceptual, academic or administrative, have badly equipped us.

One of the major problems encountered everywhere is that decisions involve both technological and policy aspects, be they political, social, industrial or cultural. This means that it would be necessary to bring together political decision-makers who are often technically ignorant and technicians who are equally often politically innocent. The difficulties in bridging this gap have been considerable. Thus, the Intelsat negotiations were in many countries looked upon as technical and operational and therefore left to the telecommunications experts who naturally dealt with issues from these points of view. However, there were important aspects which had to do with desirable patterns of communication on a world-wide level: thus, to favour technical configurations only in terms of the then current patterns of communication between industrialized countries clearly disfavoured the developing countries, both technologically and financially. Even in industrialized countries that have traditionally taken a favourable attitude towards the developing nations it was discovered late in the day that their telecommunications experts had in fact taken a stand which differed from the official foreign policy.

How then have countries fared in dealing with communications satellites? Some selected examples will have to show the type of problems encountered and the reasons for the choice of a certain course of action.

Canada and Brazil

It is interesting to compare developments in two countries which, despite their obvious differences, were faced with interesting similarities in their communications situation: Canada and Brazil. The similarities are geographical and demographic. Both countries are among the largest on earth, with a low average population density. In both countries the population has been concentrated in certain areas and in both there is a vast, partly or totally undeveloped hinterland. In Canada the population, and therefore also the communications facilities, are concentrated in an area along the southern border. In Brazil the population is similarly concentrated on a belt along the coast, from Belem to Porto Alegre. Both countries have widely scattered and isolated population groups in the interior and in both cases official policy was geared to promote those areas which were only marginally developed. A number of the reasons for interest in satellite communications were the same in both countries therefore: the possibility of getting access to a means of communication that would cover the entire territory including isolated or inaccessible areas which it would be technically difficult and economically prohibitive to reach by traditional methods.

The approaches adopted by the two countries were, however, very different. Canada showed both interest in and capacity for space activities early on. Thus, Canada became the third country after the USA and the USSR to launch a national satellite (Alouette 1 in 1962) and she has been in the foreground of satellite development ever since.

Canada's inherent dependence on communications, the success of Alouette, the achievements of the ISIS series of research satellites and the country's early participation in Intelsat were all reasons for the interest in a national satellite communications system. During the late 1960s both private and government groups undertook studies and anlyses. In 1968 the Government published a White Paper giving the general reasons which had led to the conclusions that a satellite communications system was of vital importance for the development, welfare and unity of Canada and therefore should be established as soon as possible. The White Paper gives a clear indication of the importance which was attached to satellite communications: they were seen as representing a revolution which will strongly influence future national and international communications patterns; a national Canadian satellite system was not only to improve presently available facilities but was seen as the only means of providing the rapid increase in capacity for required services.

There were different views held as to the manner in which and by whom satellite communications should be developed.

There were those who insisted that the country's telephone companies, both investor-owned and publicly-owned, should be given the mandate to add satellite-delivered signals to their terrestrial systems. There were others who opted for a government-owned and operated system, which would not be inhibited by losses likely to be incurred in the early stages of the penetration into the market place of this new technology. With that ability to choose pragmatic solutions where strict logic might decree otherwise, that often astonishes the rest of the world, the Canadian Government decided to sponsor the creation of a mixed corporation, which would have both private and public shareholdings, which would operate commercially in all essential elements but which would as well have certain 'national' objectives. [Golden, 1982.]

Thus, in 1969 Parliament established Telesat Canada with an exclusive mandate to operate a commercial satellite system. The first Anik satellite was launched already by 1972. The system has since then developed through various generations of satellites, extension of the earth station network and of the services offered. In the view of many observers, Canada has managed more successfully than most countries to design a relatively coherent policy, to provide an efficient and adequate management agency and a well-functioning operational system.

In contrast to Canada, Brazil has been through a series of interesting shifts and changes in policy and planning. Already in the late 1960s a domestic satellite system was the subject of discussions which later evolved into a fierce debate. The idea of a national satellite system dedicated to educational purposes was promoted by some educationists but mainly by the dynamic leader of the Brazilian space research organization, Fernando de Mendoça. At the space research centre in

São José dos Campos, plans were formulated for an advanced satellite system called SACI—Satélite Avançado de Comunicaões Interdisciplinárias. The original concept was for a phased development towards a national satellite system with a capacity of three television channels and 3,000 voice channels: in densely populated areas the satellite signal would be received on medium-sized earth stations for relay over normal television transmitters while in rural areas reception would be direct from the satellite on small community receivers.

Many of the ideas in the SACI project were supported by other studies, among them a joint ITU/UNESCO mission report—on condition that there was a firm policy commitment to a large-scale use of educational television and attendant changes in the educational system. The educational community was anything but agreed on the desirability of instructional television via satellite: many opposed the idea for 'pedagogical reasons'. There were also difficulties of a political and organizational nature. In the meantime, planning and construction of the national telecommunications system were given top priority in the Second National Development Plan. It seemed reasonable to install microwave links in the densely-populated coastal areas but the planned tropospheric scattering system for the interior quickly proved inadequate. The idea of a domestic system was maintained therefore, but the basic concept and direction changed from education to telecommunication. The objective of the currently conceived system is the integration of the Amazonas and the central and western regions in the national telecommunications system. The system is expected to be operational by 1985.

Southern Asia

Another interesting contrast is provided by a comparison between developments in Indonesia and India.

In Indonesia the name chosen for its satellite system, Palapa, is a symbol of national unity and integration. The government's determination to follow the capitalistic road for rapid economic development required an extensive and effective communication system for security, development administration, and for the growing business corporations and extractive industries. These factors were so critical that the question of cost and technology dependence on the West were considered problems of secondary importance to be tackled later. [Rahim, 1983, p. 57.]

Thus, television was not a primary driving force but was added at a later stage, as was the idea of using the satellite facilities for educational purposes. And only later were the attendant economic and social problems put on the agenda for a communications policy.

In contrast, Indian policy on satellite communication has been closely linked to comprehensive development policy and planning. Discussions started as early as 1960, mainly on the initiative of the well-known scientist Dr Vikram Sarabhai. A number of studies were undertaken, an institutional structure was erected through the establishment of the Indian Space Research Organization (ISRO) and the large-scale experiment in satellite instructional television known as SITE undertaken. Together, these efforts laid the foundation for Insat, a national satellite programme for telecommunications, radio, television, meteorology, remote-sensing, defence and security communications. Despite all difficulties encountered through inter-departmental rivalries and incomprehension, joint departmental planning and co-ordination have been major considerations as has the intention of achieving self-reliance in manufacturing, and institutional and managerial autonomy.

Other countries in the region have not fared so well. The Philippine domestic satellite communication programme has, in the view of one participant observer, only had modest success; he points to the need for 'rationalizing the tele-communication sector and setting the proper directives towards its development [which] has been recognized since a few decades ago. The same need is felt today except that the problems are more complex than they were before. Unless checked, the trend would be towards more complications' (Lauengco, 1983, p. 64).

To an outsider, the situation in Thailand appears confusing. Since 1979 a number of satellite projects have been proposed and approved by the government for different government agencies, including the Royal Thai Army's Supreme Command, the Communication Authority of Thailand, the Post and Telegraph Department, the Interior Minister's Office of the Under-Secretary of State, the Bangkok Entertainment Company's Channel 3 television station; and more are in the pipeline. 'Thus, while several satellite communications programs have been approved by the government and are well under way, there remains no official policy and plan on domestic satellite communications systems in Thailand.' (Supandhiloke, 1983, p. 74.)

Europe

Western Europe deserves special mention—if for nothing else than to show the difficulties of pouring new wine into old containers. Already by the end of the 1950s the lack of a European launcher rocket programme gave cause for concern. There were fears that Europe would become totally dependent on the United States for launching facilities and that European space activities would be limited to an exchange of information in the shadow of the two superpowers. The conviction grew that the Western European countries could only uphold their scientific, technological, economic and political interests through space programmes of their own. However, each nation had its own difficulties defining a coherent national policy since interests and approaches diverged widely within countries. Many of the Western European countries began their outer space activities through bilateral co-operation with the United States, through NASA's international programme. Thus, from the beginning one controversy opposed those who favoured a common European approach and those who favoured national programmes, which in most cases meant participation in the American space programme. There were also serious differences of opinion among the scientific community, those who wanted to gamble on space applications programmes and those who focussed on the perceived need for an independent European launch capability.

In the autumn of 1960 the British Government decided to cancel work on its ballistic missile Blue Streak. In order to maintain its space industries and not totally waste the £80 million that had already been spent on Blue Streak, the

United Kingdom proposed to other European governments a common European rocket programme based on Blue Streak. France was the first country to back the British proposal and suggested the use of the French Coralie rocket as the second stage of the joint launcher, while other tasks would be parcelled out among other European countries. Thus, in 1962 was created the first European space organization, under the name European Launcher Development Organization, ELDO. However, ELDO only grouped Belgium, France, Italy, the Netherlands and the United Kingdom and West Germany; the other European countries did not adhere, for political and economic reasons.

The creation of ELDO was little more than an enthusiastic and some-what naive response to a suggestion for a European rocket . . . It was designed originally to become the basis of a joint satellite launching programme, but little or no attention was paid to the sort of satellite the Europe 1 (as it came to be named) would actually carry into space. [Schwarz, 1981, p. 66.]

There were other difficulties and controversies: over the financial contributions to ELDO, over the procurement policy, over the level of participation by each member, over escalating costs. ELDO reeled from crisis to crisis: in 1966 and in 1968 there were two major crises which were barely overcome. The question of Europe's capability for an independent launcher divided the European space nations. The United Kingdom wanted to forget the whole business of a European rocket and, based on its special relationship with the United States, claimed that launchers would probably be available on acceptable terms and at lower prices from the States. France, which has always been the most ambitious space nation in Western Europe, with the support of West Germany, Italy, the Netherlands and Belgium, believed it was absolutely essential for Europe to be autonomous in space.

In the meantime, European co-operation had begun in the scientific domain and the result was another organization, the European Space Research Organization, ESRO, which was founded in 1964. All of the ELDO countries became members of ESRO, with the addition of Denmark, Spain, Sweden and Switzerland.

Despite some initial success, ESRO also and quite quickly got into deep water. Opinions were divided on all important questions: should ESRO focus on sounding rockets and smaller scientific experiments or concentrate on large projects like a space laboratory? Should the main focus be in scientific experiments or should ESRO also, and even primarily, develop applications satellites in such areas as meteorology and communications? And as in ELDO, there were controversies over the budget and the allocation of contracts: since national governments contributed to ESRO according to their gross national products, the problem was how to return the same proportion of money to the national industries. Thus, when proposals for major contracts came up, national delegations would vote according to the share their country could expect in return, rather than on technical competence.

There were, however, voices raised against the prevailing confusion. The Council of Europe had already in 1960 proposed a joint European space programme, though to no effect. More reports were produced and in 1966 Lord Jellicoe succinctly put the basic questions anew:

— Will Europe remain in space or withdraw because of too high cost?
— Will Europe continue to develop a rocket capability?
— Will Europe develop an independent capability with regard to communication satellites?
— Will Europe develop satellites for other peaceful purposes?
— In view of the above, what would be a reasonable budget for a common European space programme?
— Which would be the best institutional structure for a future European space programme?
— What will be the importance of co-operation between Europe and other regions in the world? [Council of Europe, 1966.]

There are obvious reasons for asking all these questions. In 1966 it had become obvious that there was a need to co-ordinate all the European space organizations and so another organization was added: the European Space Con-ference. In the meantime, European industry had combined in an organization called Eurospace which also fought for one European space organization and for a co-ordinated space programme. There were also the regional organizations on the telecommunications side (Conférence Européenne des Postes et Télécommunications, CEPT) and on the broad-casting side (European Broadcasting Union, EBU) which had a strong interest in the European plans—since they were supposed to be the users of future communication satellite systems. Thus, ESRO was supposed to develop the satellites, Eurospace to manufacture them, ELDO to launch them, CETS to co-ordinate views while CEPT and EBU were supposed to use them. However, no one had the mandate to co-ordinate activities of these various bodies. The complex international situation appears simple compared to this European confusion. Sometimes it seemed almost impossible to find one's way through the wild flora of organizations, committees with sub-committees and their sub-groups, conferences, crises, resolutions and studies with high-sounding names: 'Economic Potential for Europe of Application Satellites', 'European Satellite Programme for Telephony and Television Distribution'. The comment by an American observer seemed pertinent: 'Each time the Americans launch a new satellite, the Europeans establish a new committee'.

If by 1968 it was recognized that the first phase of Euro-pean collaboration in space technology had been a failure, the next period was not much better. In each of the major countries—France, the United Kingdom and West Germany —the situation has been described as characterized by political delays and continuous incapacity to reach agree-ment which stopped just short of total breakdown. In 1973 the situation was rescued by a package deal which seemed to ensure the future of European space co-operation but which, in the opinion of some observers, really accepted defeat for the ideals of European unity. It is true that the members of ESRO agreed on the establishment of a new and unified organization, the European Space Agency, ESA, which would replace both ELDO and ESRO. However, the agreed package deal also implied that a country's preferences for particular

bits of joint projects could be reflected in its financial contributions and its expectations of industrial contracts.

The major ESA development programmes are the Ariane
rocket launcher, mainly paid for by France, the Spacelab,
with Germany as the largest contributor, and the communications satellite programme which seems to have been overtaken by individual countries deciding to proceed alone (UK),
bilaterally (France and Germany), or even sub-regionally,
as planned by the Nordic countries.

The main task of ESA was to develop a coherent space
policy for the member countries. By and large it faced the
same problems as its predecessors. However, the main problem in the past had been political: to reach a common policy
it was necessary to find agreement between governments who
not only had different views and hidden agendas but sometimes lacked a coherent national space policy altogether.
Recently, the problem of differing national interests has been
intensified as countries have become more aware of the
commercial prospects of space activities, particularly in
communications. However, evaluations of ESA vary. At
UNISPACE 82, ESA was presented as an example of regional
pooling of resources to achieve objectives that are beyond the
reach of individual states, an example which would also be
adopted by groups of developing countries. Other observers
see ESA in a bleaker light:

As space becomes another area of intense economic competition the
political will to collaborate within a regional organization such as the
ESA rapidly fades away. ESA's painfully slow process of arriving at
politically acceptable decisions is being overtaken by the faster process
of commercial expansion . . . When [if] Europe finally wakes up
to the reality of the politics and economics of space technology, it
may do so too late to keep the common idea alive. In the beginning,
there was a great willingness to share resources and efforts in order
to build up a technological capability. Now that European space activities have matured, the road towards great European integration is
rapidly being closed. [Schwarz, 1981, p. 71.]

Thus, despite the fact that satellite communications have
an inherently international dimension, systems have generally
developed on the basis of national objectives and programmes.
In so far as smaller countries cannot agree on common

undertakings, the dominant position of the major space powers increases. However, the possible overcrowding of the geo-stationary satellite orbit and the prospect of multi-purpose space platforms, antenna farms and other technical advances designed to provide a more economic use of the orbit and the radio frequency spectrum might make new forms of international co-operation and joint planning necessary.

7   Policy and law in the making

As should be evident from the previous chapters, satellite communications raises novel issues in a great number of fields. Each of these issues is complex enough to require special treatment for any in-depth analysis. In this chapter the focus will be not only on the 'what' of selected major issues but also on the 'how': how have individual countries and the international community tried to go about solving the emerging problems, in so far as they have managed to define them in the first place. Thus, the focus will be on policy and law: the making of policy and of law and on the development of institutional structures for these purposes.

Basic issues

In present-day mythology, one cherished idea is that policy-making and planning are rational processes, based on logic, coherence and a fair input of relevant information. Most writing on these matters seems to share this assumption, which lends it an air of unreality. If nothing else, a look at the results of these activities should be enough to prove how far from the reality is this ideal picture. Rather, one would be justified in feeling that even to discuss policy-making in such a complex field as satellite communications is enough to stir up a hornets' nest: the buzz of unresolved problems, contradictory or divisive attitudes, opposing ideologies might appear threatening. However, in certain respects the policy aspects of satellite communications seem to be the most important of all, since they are placed at the intersection of two major issue and policy areas: satellite communications are uneasily suspended between outer space issues and issues of communications and information, ranging from monitoring of the earth via telecommunications, to the media. The problems in these two areas are similar to those in other new areas of public decision-making such as the environment, energy or ocean resources. They cut across traditional administrative, legal and intellectual categories and demand an inter-disciplinary and inter-departmental approach. Thus, the categories, the 'mental boxes' which condition our current institutional and legal structures appear outdated and inadequate. The initial common problem is therefore that individual countries and the international community not only have to evolve adequate policies of a new kind but that, at the same time, they have to work out an adequate policy-making machinery, to develop social structures that correspond to the nature of the issues to be solved.

The second problem refers to the difficulties in fitting policies for satellite communications into the policy patterns of other related areas such as the media. At present much attention is lavished on policies and regulation for different branches and media of communication. The assumption seems to be that, together, these measures make up a 'communications policy'. However, if policy is understood to mean a comprehensive approach to a coherent course of action, then this aggregate provides policy by default only. At present most 'communications policy' consists of reactions to technical and social pressures. Generally there is no one policy but a series of policies formulated and implemented by a diversity of public and private agencies in pursuit of limited and often contradictory objectives. Present work towards national communications policies often looks like an attempt at putting together a large jigsaw puzzle without having decided how many or which pieces should be used nor what the final picture should look like. We have not even been able to establish a relevant framework: we are uneasy about what 'policy' should be taken to mean with reference to communications; we do not appear to be capable of defining 'communications' as an object of policy nor have we analysed the scope of action possible under the heading 'national'. Matters become even more complicated when the need arises to co-ordinate with policies in other areas such as education which is in the grip of its own policy crisis. Educational television has proved difficult enough to handle: educational television via satellites compounds the problems.

As stated by one expert with reference to the American situation:

A new framework for understanding the international sphere of communications and information (C&I) policy is needed. The old separation into neat compartments such as domestic vs. international, economic vs. political, transmission vs. content, civilian vs. military, strategic vs. civilian trade, domestic vs. multinational corporations vs. foreign firms, national security vs. civil liberties, government-conducted foreign policy vs. media diplomacy, military vs. economic intelligence, and privacy vs. freedom of information, can no longer be sustained. [Ganley, 1980, p. 1.]

Obviously, extreme regulation and anarchy are equally undesirable, but in the absence of even an adequate conceptual framework we seem unable to define which levels of order and disorder we can feel comfortable with.

A third problem concerns the definition of criteria on which to base priorities. In a very simplified form: for which purposes is satellite communications more important and which is the relative priority to be given to: telephony, broadcasting, navigation, meteorology or systems for the collection of data on resources, pollution, crops and water? Each communications and information service raises its own imperative demands which have to be weighed against another. In addition, new technologies and services give rise to new interest and pressure groups which clamour for their share of available resources and for supporting regulation.

We still deal with such questions on an *ad hoc* or piecemeal basis conditioned by a short-term perspective. In a larger perspective the problems we face concern the choice between different systems for the transport of people, goods, services and ideas. And behind all these separate problems lies a basic issue: on which criteria and policies should we define the kind of communications systems and information flows that are required for changing social functions, which rules and practices respond to the aspirations of different social groups, to the overall needs of national societies and the international community?

A further cause of present difficulties is that concepts, policies and rules in the communications and information field have generally been linked to a particular technology or level of technology. Policy and structural frameworks antedate the communications revolution. Thus we separate the 'medium' from the 'message' and deal differently with information on physical support and electronically transmitted information. Policies and rules that have been evolved for one technical mode of communication are stretched beyond their inherent capacity to cover new situations. This approach becomes particularly inadequate in a situation where technologies and systems tend to converge in new, unforeseen combinations.

It is therefore not surprising that attitudes towards such new communications technologies as satellites are contradictory. Euphoric, technology-based forecasts provide no more guidance than equally exaggerated anxieties. The evaluation of the social consequences of new communications systems is no less confused: some fear an electronic Big Brother or the Global Village or corporate media imperialism while others, with equal fear, foresee a lack of social cohesion through utter fragmentation of audiences and producing entities.

To a large extent the problems currently experienced in defining national policies depend on very real difficulties in dealing with the interrelationship between communications and other social processes and products. To escape this dilemma, countries have focused on specific communications issues in isolation, almost in a vacuum, without clear reference to the overall societal context. However, the space dimension introduced by satellite communications has often made it necessary to widen the view and to look at a larger social landscape. Therefore, it is useful to mention some of the underlying factors that condition policy-making, structures and regulations as regards satellite communications.

Factors in national policy-making

The first and most obvious factor influencing policy and regulation in the communications–information field is the socio-economic ideology and structure of a given society. In relation to communications, a distinction can be made between:

(i) Countries adhering to a 'liberal', free market economic philosophy and structure where the general approach is for communications to be conducted, as far as possible, in and by the private sector; the most extreme example is the United States.

(ii) Countries with a mixed economic structure where one of the major issues concerns the balance to be struck between the private and public sector, between official planning and individual, group or private enterprise initiative; to this group would belong the Western European countries and Canada, Australia and Japan.

(iii) Countries with a 'socialist' ideology and a planned economy where the public sector also dominates in communications. To this group belong the USSR and to varying degrees the Eastern European countries.

In this perspective the developing countries, despite the obvious and profound differences in situations and systems, share one characteristic: in the case of modern communications the models applied have generally been imposed or imported from outside and therefore a decisive aspect of their policy-making is to evolve new structures and rules which correspond to their specific conditions and requirements.

In many situations a major issue is the way in which communications and information policy revolve around the private/public sector axis. The interesting point is that while there are great differences between countries in the three groups with regard to policies in some areas (i.e. the media) there are great similarities in other areas. Thus, in all countries the mail has become a public monopoly. Telecommunications and broadcasting show a more diversified picture but, whatever the system adopted, there is one aspect which in all countries is the responsibility of central government: the regulation and use of the radio frequencies. There was from the beginning therefore one aspect of satellite communications for which policies had to develop in the public sector. The socialist countries did not have any difficulties with other aspects: satellite communications were firmly and logically placed in the public sector as a state activity.

The mixed economy countries seemed to regard satellite communications as too important in relation to public affairs to be given over to the private sector. Various forms were used: satellite communications were put under a special authority or placed under existing structures but all were public or semi-public. Even the United States had difficulties in giving satellite communication totally to the private sector: government regulation is exercised through the FCC and certain government departments; NASA has been deeply involved in the development of communications satellites and even Comsat is subject to government supervision in certain respects. One can, however, view the history of communication satellites in the United States as the constant struggle by the private sector to take over this new technology, a trend which has been greatly assisted by the policy of 'de-regulation' in the communications field pursued by recent administrations.

It is striking that few, if any, countries seem to have evolved adequate decision-making structures in this area. At present the most obvious feature is the dispersal of function and responsibility among different authorities and agencies. However, despite this dispersal of social mandates and despite the variety of political and socio-economic systems, the formal institutional arrangements adopted in the countries of the world can be fitted within a surprisingly limited number of categories. In general there appear to be three models in use:

(i) Communications and information policy is left to existing institutions and carried out on a sectoral basis; this model is the usual one in both industrialized and developing countries and implies that responsibility is dispersed among traditionally defined ministries and other public authorities responsible for telecommunications, various media, intellectual property rights, trade and foreign policy, etc.

(ii) Some countries have responded to these problems by setting up permanent, specialized policy- and/or decision-making bodies. There are two main versions of this model: permanent bodies with decision-making powers such as the Federal Communications Commission, USA, the Canadian Radio, Television and Telecommunications Commission, the recently created Haute Autorité de l'Audio-Visuel in France; and permanent bodies with only advisory functions ('communications' or 'media' councils in a number of countries). Also, the mandates given such bodies show great variations: some combine telecommunications and broadcasting which may or may not include cable television, others deal with all the audio-visual media, others only with data protection, etc.

(iii) However, many countries seem to find that neither of these structures is adequate since, in addition, they have recourse to temporary *ad hoc* commissions of inquiry set up to study and propose policy for specific issues. During the last few years there has been an extraordinary proliferation of such *ad hoc* committees in all parts of the world. Again, though, the response within countries to technological and social pressures often seems surprisingly dispersed. During the last few years, committees of inquiry in a country such as the United Kingdom have investigated the future of broadcasting, the film industry, the press, data protection, the organization of the Post Office, revision of the copyright legislation and a national communications satellite system without any visible co-ordinating machinery. It is revealing though that satellite communications have been a choice subject for such *ad hoc* inquiries. There seem to be few countries concerned that have not conducted their own inquiries into the major issues of communications via satellites.

Of particular relevance would therefore be the attempts to formulate the bases for policy through inquiries and studies of a more comprehensive kind. Interestingly enough, most of these efforts have been initiated from the telecommunications side. In this respect, the Canadian Telecommission (1969–71) still represents the most thorough, open-minded and lively investigation into modern communications undertaken anywhere. More than forty separate studies were grouped under eight main headings: legal considerations, economic considerations, international issues, technological studies, information and data systems, telecommunications environment, telecommunications and government, special studies. Unfortunately, for administrative reasons, broadcasting not being part of the mandate of the Department of Communications, it was not included in the Telecommission's brief; however, the report published under the title 'Instant World' is wide-ranging enough to include such aspects as telecommunications and the arts. The Telecommission and its report remain an exemplary but unfortunately rare attempt to provide a coherent basis both for public debate and policy formation.

Among more recent, interesting examples figures the Committee on the Future of the Telecommunications System in West Germany and, particularly, the report 'Telecom 2000' prepared by the National Telecommunications Planning Branch of the Australian Telecommunications Commission, in 1976. 'Telecom 2000' is unique in the sense that the report begins with studies of the probable and desirable future development of Australian society. These studies provide the background and framework for the analysis of what technical systems might be available and how they could fit into various predictions. Great attention is paid to the use of telecommunications as a means of implementing stated social and economic policies: economic growth, resource conservation, decentralization, growth of knowledge industries, and encouragement of 'open government' and community participation.

The approach adopted by the Swedish Secretariat for Future Studies was very different, as reflected in the overall title given to series of studies and other activities in the early and mid-1970s: 'Man in the communications society of the future'. An unusual feature was the emphasis on the individual, on the bio-psychological and other conditions affecting the information-handling capacity of human beings.

In recent years more comprehensive analyses have been attempted in a growing number of countries and from more varied perspectives. There is the grand new mode in economic analysis which began in 1962 with the publication of Fritz Machlup's study on 'The production and distribution of knowledge in the United States', and was followed by studies on the economic importance of a widely defined 'information sector' representing the aggregated information activities in society. Some economists have even replaced the traditional concepts of capital and labour by energy and information as basic conceptual objects (Attali). In Japan, both studies and experiments concerned a great variety of aspects of the emerging 'information society'.

Other studies have focused on the importance of the telecommunications and information sector for national sovereignty and on the impact of the increasing 'computerization' of society. The perceived vulnerability of the computerized society was given increasing attention in countries such as France and Sweden, followed in France by a series of studies and decisions concerning the audio-visual sector, combining industrial, cultural and communications policies. At the international level, the non-aligned and other developing countries focused on various aspects of the issues subsumed under the concept of a new world information and communications order, as did the UNESCO-sponsored McBride Commission which published its much-discussed findings under the title 'One world—many voices'.

Also, in time for World Communications Year 1983, the ITU presented the results of a study undertaken jointly with the OECD under the title 'Telecommunications and Development'. This report, which represents the result of pioneering research, set out to produce evidence on a number of misconceptions, such as the notion that telephone systems are of primary interest to urban areas and to the wealthier groups in society but do not contribute to raising the living standards of the population as a whole, particularly in rural areas; or the view that investment in telecommunications does not warrant the same priority as other categories of public infrastructure such as roads, energy, irrigation or health services. The study reached the conclusion that the telephone seems to be a far more important factor in the development process than previously thought. The report also gives economic evaluations of the GLODOM plan, which is designed to provide:

A global approach to the major telecommunications challenge of the decade, namely, how to bring low-cost domestic telecommunication services to support the integrated development of rural areas, which are inhabited by the significant majority of the world's population presently denied such services. [Butler, 1983, p. 4.]

Thus, the report emphasizes the advantages of suitable mixes of modern space and terrestrial technologies to meet these needs. The report will also provide an input to the Independent International Commission for World-Wide Telecommunications Development which was set up by the ITU in late 1982.

While all these studies were proceeding—and regardless of their value—countries had to take decisions, often before they were ready, under pressure from outside developments. The manner in which governments have approached the advent of satellite communications is therefore revealing. In many countries policy-making in this new area was simply divided according to traditional categories and grafted on to existing structures, such as ministries and other agencies responsible for telecommunications, industrial development, the media, meteorology, foreign policy, etc. It has been stated that in the United Kingdom space policy decisions at one stage required the concurrence of some fifty-nine agencies. The necessary co-ordination is cumbersome, goes against the bureaucratic imperative and is therefore often used only in reaction to urgent needs or crises but not as a toll for long-term policy. The result is often no policy at all, or a series of policies that suffer from frequent changes, built-in contradictions or other difficulties.

In some countries a mixture of policy and operational mandates was given to newly-created agencies specifically responsible for space matters; examples are NASA in the United States, ISRO in India and NASDA in Japan. Overall, these countries seem to have managed their space policies better than the others. A third variation has resulted in the creation of special entities responsible for satellite communications, for example Telesat (Canada), Comsat in the United States and Telespazio in Italy. Again it seems that such a structure has helped these countries in the formulation of satellite policy.

International linkages

There are today few areas where national and international policy-making are not interlinked. This is as true of outer space as it is of issues in the communications and information field, thus doubly so of satellite communications. The linkage between domestic and international policy-making is in turn closely related to another typical feature of international relations: the diversity of 'actors' operating at the international level. Beyond the state—or beside or below—now function a number of other actors which have so changed the scene that an entire new branch of international relations study has arisen under such names as 'transnational studies'.

The phenomenon of transnational actors is in itself anything but new: for centuries the Vatican has been—and still remains—one of the most important non-state, transnational actors on the international scene. What is new is the increase in the number, variety and functions of transnational actors. They now range from multinational companies to extra-territorial terrorists who are not territorially bound but have established sets of relationships within and across state boundaries, and present new challenges to states. In every field of human activity, new groupings and new linkages between national groups and interests have transformed the international scene: in business, religion, sport, entertainment, science, crime, the arts and the professions. The processes represented by these new actors are reflected in the phenomenal increase in non-governmental organizations and the equally astounding number of international conferences, meetings and symposia. These various interests and groupings form criss-crossing alliances across national borders, working sometimes for, sometimes against, the policies of their governments who thus have to contend not only with the competing domestic constituencies but also with the competing interests of the transnational groups.

These developments are clearly reflected in the field of satellite communications where one has to include the transnational actors in both the space area and in the domain of communications and information. The interests involved range from the aerospace industry to copyright owners, and include telecommunications agencies, banks and insurance companies, data bureaux and broadcasters, radio astronomers and environmentalists, as well as the various tiers of intergovernmental organizations.

While the formal responses of the international community are spread throughout the loose system of intergovernmental organizations, another major difficulty in trying to make sense of the confusing international scene is the absence of any generally agreed conceptual approach to account for the variety of actors and their relationship across, beyond, or below the state.

Levels of institutionalization

However, some attempts have been made to analyse the various levels of institutionalization for collective responses within the international community. One of these schemes was developed particularly in respect of new situations caused by advances in science and technology and ought therefore to be relevant to satellite communications.

Despite the somewhat awesome terminology, the basic thrust of this analysis is relatively straightforward. As proposed by the American scholar J. G. Ruggie, a distinction is made between three levels of institutionalization: cognitive or epistemic communities, international regimes and international organization.

(i) The first level is represented by what Ruggie (1975) defines as 'epistemic' or 'cognitive' communities. This concept is based on what Ruggie, borrowing from the French structuralist Michel Foucault, calls the 'epistemes' through which political relationships are perceived. By 'episteme' Foucault means a dominant way of looking at reality, a set of shared symbols and references, mutual expectations. Epistemic or cognitive communities are thus said to grow up around shared epistemes which define and delimit for the members the proper construction of reality. Such epistemic communities may derive from bureaucratic positions, technical training, scientific outlook or shared disciplinary background, or from such functions as representing national public authorities at the international level.

In the satellite field this level of institutionalization is a characteristic feature. The epistemic communities are abundant and involve both governmental and other actors. They form criss-crossing patterns of relationships which are sometimes heavily organized, sometimes fluid. Often, they appear isolated from one another although they may combine in alliance on specific issues.

These communities can also be defined as sets of experts whether involved in inter-governmental or non-governmental organizations, whether representing national authorities or non-state interests. To a large extent their activities can be seen to focus on or crystallize around specific inter-governmental organizations where the final decisions will be made and regimes defined.

In relation to satellite communications, these communities or sets of experts include:

(a) First, a space-satellite community which mainly comprises the aerospace industry and related industrial sectors, primarily in electronics, operational agencies at the national, regional or international level (e.g. Telespazio (Italy), Telesat (Canada), Indian Space Research Organization, Arabsat, Intelsat, Intersputnik, etc.) and specialized research institutions; within this community there are subsets concerned with remote sensing, telecommunications satellites, etc.

(b) A set of governmental, diplomatic-legal experts who mainly deal with the juridico-political issues in the context of the United Nations and its Committee on the Peaceful Uses of Outer Space.

(c) A set of technical-regulatory experts, primarily concerned with telecommunications, whose work and contacts are centred on the International Telecommunication Union and regional telecommunications agencies.

(d) Various user communities are represented by sets of experts in the mass media field (press, news agencies, broadcasting), data transmission, meteorology, navigation, etc.

(e) Last but not least, the epistemic community represented by the military.

The interaction between such communities or sets of experts concerned with satellite communications varies with the issues involved but is often surprisingly weak. This reflects the prevailing conceptual and administrative compartmentalization at the national and international level. Thus, it was only the complexity of issues, the variety of expertise required and of interests involved in the elaboration of legal principles for satellite broadcasting that for the first time brought together the diplomatic-legal experts, the technical-regulatory community, the space satellite set and the broadcasting sub-set. However, it took quite some time before the representatives of these various communities had managed to find a minimum common ground for their discussions since, to start with, they did not even share a common terminology. Similarly, the increasing economic and political importance of telecommunications, including space communications, brought the diplomatic-legal set into the confines of the ITU—to the great consternation of the technical experts who bewailed the intrusion of 'politics' into their 'pure' technical domain.

(ii) The next level of international institutionalization is represented by international regimes. The term regime in this context is defined to mean sets of generally agreed rules and regulations, plans and organizational energies as well as financial commitments accepted by all or groups of the parties involved. Used in this specific sense, regimes can be exemplified by the international system of safeguarding nuclear materials or by the World Weather Watch mentioned earlier which comprises 'national weather bureaux doing what they have always done, doing some things they have never done and doing some things differently than in the past, all in accordance with a collectively defined and agreed-to plan and implementation programme.' (Ruggie, 1975, p. 571.)

International regimes may be differentiated by the purposes they serve, the instrumentalities they use and the functions these instrumentalities actually perform. Applied to the satellite communications field, this concept of international regimes is helpful in revealing the absence of any single overall regime. Instead, there has emerged a number of partly over-lapping, often unco-ordinated and sometimes contradictory or competing regimes. Furthermore, satellite communication is subject to the influence or application of regimes evolving in other areas which directly or indirectly relate or are made to relate to the satellite regimes.

In the satellite communications field, the goal formulated in the United Nations of developing a common framework for national and international behaviour has so far met with some, if limited, success. The Outer Space Treaty of 1967 and subsequent legal agreements represent a major achievement although the pressure on this legal edifice is more than it can support. The technical-regulatory regime developed in the context of the ITU has been successful but will probably require reinforcement to withstand increased tension and pressure. Characteristic for satellite communications is the development of specific regimes for the acquisition of joint facilities—organizational, financial, operational—through such organizations as Intelsat, Intersputnik, Inmarsat, Arabsat and the European Space Agency.

(iii) In this approach, international organization represents the most concrete and best defined level of institutionaliza-tion. The immediate environment of international organiza-tions consists of the regimes they serve. Thus, any international organization may be seen as operating in a three-dimensional policy space whose axes are defined by the selected features of the surrounding regime, i.e., in this model: purposes, instrumentalities and functions. Also, their tasks are seen in a three-dimensional perspective. They may perform a facilita-tive task when planning for the regime is carried out within the organization but decision-making and implementation are left to its members. When both planning and decision-making are carried out within the organization and imple-mentation by its members, the task is described as enabling. When all three activities—planning, decision-making and implementation—are carried out by the organization its task is defined as operational. There is a fourth case which should be added to this scheme: when planning and implementation are carried out by the organization, in co-operation with the members, but decision-making is left to the members, there is a so-far unnamed task which we might call quasi-operational.

This typology, which differs in many respects from other attempts to categorize international organizations, has the advantage of clarifying the relationship between the organization and its members. In its original form, this scheme was applied to inter-governmental organizations but it would in fact be valid for many non-govermental organizations as well.

Most organizations concerned with satellite questions, and particularly those of a multi-purpose character,would show a mix of tasks. Thus, in the current situation, inter-governmental organizations have a facilitative task with varying degrees of enabling and quasi-operational tasks. This, in principle, would seem to be true of the United Nations system of organizations and most regional inter-governmental organizations (such as OAU, Council of Europe, etc.). The pure operational case is rare but is represented in the satellite field by such new organizational types as Intelsat and Inmarsat. A non-governmental professional organization such as the European Broadcasting Union represents a variation of the quasi-operational type. With regard to, say, news exchanges over a network that includes satellites, the principles of operation are drawn up within the organization, the implementation is carried out by both the organization and its members while the decision to offer or receive news items is left to the members. This schematic model is useful as background to a rough map of the complex—and complicated—international structure at this third level of institutionalization.

International structures

A characteristic feature of the international policy-making and law-making structure concerned with satellite communications is the lack of unity and coherence. It would be more appropriate to speak not of structure but of structures, not of one regime but of regimes, and thus of dispersal of function and responsibility. In general terms the sectoral approach, still prevalent at the national level, conditions decision-making processes at the international level. Thus, as indicated earlier, in respect of the epistemic communities involved, telecommunications policy and law, including rules for the conduct of satellite communications, will be pursued in one forum (ITU), different in approach, mandate and practice from those where the juridico-political aspects, questions concerning the media, copyright or trade aspects are handled.

Various approaches are possible to provide an overview of the structural framework within which the processes of developing principles and rules for satellite communications take place. In this context it has seemed preferable to use an organizational approach as a more easily comprehended framework for the present distribution of policy-making and law-making functions.

(i) *The United Nations*

Within the United Nations system a distinction can be made between the specialized agencies which have become involved in satellite communications and will be mentioned below and the activities carried out by the United Nations itself mainly through the General Assembly and directly related organs.

United Nations concern with satellite communications issues may be seen under three aspects:

— development of international law: of specific importance in this context is the development of space law through the Outer Space Committee and its organs, in particular the Legal Sub-Committee;
— standard setting with regard to human rights: the most relevant aspects concern the principles on freedom of information which have been evoked with reference to satellite broadcasting;
— economic and social development including access by developing countries to space technology.

The specialized agencies within the United Nations system that are directly involved with satellite communications include first of all the International Telecommunication Union which provides a rule-making function particularly with regard to the allocation of radio frequencies, among services and countries, regulation of the use of the radio frequency spectrum and the geo-stationary orbit and the establishment of technical standards.

Such agencies as the International Civil Aviation Organization (ICAO), the Intergovernmental Maritime Organization (IMO), the World Meteorological Organization (WMO) represent important user groups and user interests whose requirements can play an important role in the allocation of frequencies for the different satellite services. UNESCO for its part is involved in the promotion of the use of satellite communications for educational purposes, and for the exchange of information, both in terms of news and of scientific and technical data.

(ii) *Regional inter-governmental organizations*

At the regional level a number of organizations have concerned themselves with various aspects of satellite communications. In Western Europe, as a first example, these include the Council of Europe, particularly in respect of implications in such areas as education, culture and the mass media. In relation to industrial development and certain information issues the European Economic Community (EEC) has recently paid increasing attention to satellite communications, specifically satellite broadcasting; the same is true of the Nordic Council which has initiated the concept of a common Nordic satellite broadcasting system.

The same trend towards increasing attention on satellite communications can be seen in other regions. Thus, such organizations and their associated, specialized bodies as the Andean Pact (Pacto Andino), the League of Arab States, the Association of South-East Asian Nations (ASEAN) have taken various steps with regard to plans for the use or establishment of satellite systems at the national and/or regional level.

(iii) *Operational agencies*

The inter-governmental organizations mentioned so far are mainly concerned with policy-formation and standard-setting, and in certain cases, the promotion of operational agreements. As far as possible, new communications technologies and services are incorporated in the mandates of already existing institutions such as the regional telecommunications organizations. However, satellites, because of their special characteristics, have spawned a series of new operational organizations at the international as well as at the regional level (Intelsat, Intersputnik, Inmarsat, Eutelsat, Arabsat, etc.).

(iv) *Professional and trade organizations*

The increasing density of international relations is also reflected in the proliferation of professional and trade organizations in all areas. In the satellite field it is, in particular, various user organizations that play an important role, such as the regional broadcasting unions, or the International Press Telecommunication Council (IPTC).

Legal framework

This very rough sketch only gives an approximate image of the complex, sometimes almost impenetrable, web of organizations with varying mandates and functions that are involved in satellite communications, directly or indirectly. In very general terms, it can be said that each one of these organizations is related to one or several specific legal regimes developed within a particular organizational environment. An indication will be given below of the different legal areas that from different points of view cover matters related to satellite communications. However, neither national legislation nor the international legal framework provide for a coherent 'communications' or 'information' law. Such law as exists is pluralistic, unco-ordinated and mostly based on limited, functional objectives. Communications and information are the concern of various branches of law which are of varied origin and separate evolution, drawing upon different concepts and legal approaches, and resulting in legal rules that are, to varying degrees, deficient and often contradictory. Since there is at present no generally agreed conceptual approach in this area, the framework used is one that was recently developed by the author to cover all branches of law dealing with communications and information (Ploman, 1982). In this context only those legal regimes specifically relevant to communications satellites will be mentioned.

However, it should first be noted that general rules of international law apply to state behaviour and international relations in the field of satellite communications as they do in all other areas. Thus, the general principles laid down in such binding international legal instruments as the United Nations Charter are applicable in this field. Also, rules which affect satellite communications and their uses may be included in international instruments covering other subjects (such as human rights conventions, trade agreements, etc.). The branches of international law that specifically deal with subject matters relevant to satellite communications include the following:

(i) Information law, which is here used to designate legal rules and regulations referring to freedom of information, and the related protection of individual rights (privacy, etc.). It should be noted that, at the international level, such traditional concepts as freedom of information and free flow of information tend to be complemented by new concepts such as the new international information order, the right to information and the right to communicate.

(ii) Media law in this context refers to the regulation of specific media or means of communication. In general, particular media are not specifically mentioned in international law where more general and vague expressions are used such as 'information media' or 'media of mass communications'. However, an International Convention on the Use of Broadcasting for Peace of 1936 is still in force; the European Convention on Human Rights includes specific reference to the regulation of cinema and broadcasting, and in the United Nations much effort has been spent on trying to reach agreement on the legal principles for television broadcasting via satellites.

(iii) Telecommunications law includes the technical-administrative regulation of telecommunications, including broadcasting and other special services. Not only the International Telecommunication Convention but also the detailed Administrative Regulations— for telegraph, telephone and for radio communications —are binding on all member-states of the ITU. The Convention and the Regulations also include a number of important general principles of importance for satellite communications, e.g. equal access by all countries to the radio frequency spectrum and the geostationary orbit, international protection—on certain conditions—of frequency assignments, etc.

(iv) The new branch of international law dealing with outer space has found its main expression in the Outer Space Treaty of 1967 (the full name is Treaty on Principles Governing the Activities of States in the Exploration and Use of Outer Space, including the Moon and Other Celestial Bodies). Space law is formulated mainly under the auspices of the United Nations through the Outer Space Committee which has followed the conclusion of the Outer Space Treaty by a series of agreements covering specific issues such as the liability for outer space activities, the rescue of astronauts and the exploration and use of the moon. The Committee has for a number of years been working on agreements in two areas of satellite communications: international legal principles for television broadcasting via satellites and for remote sensing via satellites.

(v) Intellectual property rights might seem far removed from satellite communications, but there were early attempts to provide for special copyright rules in respect of the use of satellites. This approach, however, was rightly shelved. In contrast, the special issue of the protection of television signals transmitted over communications (but not broadcasting) satellites was regarded as falling uneasily between telecommunication law and copyright law; a special convention on this subject was concluded in 1975.

(vi) The notion of informatics law is here used to refer to the specific regulation governing computerized information services and transborder data flows. The issues which so far have retained the attention of international law-making organizations concern the consequences of technical developments for the exercise and protection of human rights: areas for possible agreement have been discussed in the United Nations, while at the regional level a convention for the protection of privacy has been concluded within the framework of the Council of Europe. There are also proposals to provide for a more general regulation of transborder data flows; however, these aspects have not yet been clearly formulated and represent a new, and still emerging legal area.

(vii) Trade and customs regulation. For the movement of information in material form, customs agreements and rules are applicable but they have so far not been extended to the electronic transfer of information. However, the international movement of computerized information in the form of data which to a large extent are carried by satellite increasingly presents important trade and economic aspects. Agreements on trade and economic relations can therefore be expected to play a growing part with regard to information carried by satellites.

(viii) Law pertaining to education and culture may be regarded as a vague, still emerging aspect of the law. It finds its source in the explicit statement of the right to education and to culture included in all major human rights instruments. Most international agreements in this area are non-binding but do often refer to the use of new technologies for the achievement of certain stated objectives in the areas of education and culture.

(ix) Similar to the numerous international agreements in other fields, those concluded with reference to collective security and military alliances include provisions concerning communications, and also satellite communication. Thus, there exists within the framework of NATO an international agreement on the use of communications satellites.

Issues and developments

Policy-making structures and legal regimes are subject to three major influences:

— the continuing transformation of international relations and of the international system;
— changes in attitude towards the communications–information complex and the attendant location of issues in this field high on the political agenda;
— the consequences of the rapid development and introduction of new communications and information technologies and services.

These three dimensions to a large extent underlie analyses right through this book; they are mentioned here specifically as a background to an account of some reasons for the lack of coherence and consistency in international legal concepts and applicable provisions. The first reason has already been mentioned. Generally our current concepts, structures and legal regimes in the communications field suffer from the constraint of having been attached to a particular technology or level of technology. Policy and legal regimes mostly antedate the communications revolution. We still separate the 'medium' from the 'message' and thus deal differently with the same information, whether in material form (e.g. film), or transmitted electronically (for example, in terms of customs regulation). Rules that have been evolved for one technical mode of communication are stretched beyond their inherent capacity to cover new requirements. Thus, when new technologies and services are added to the existing ones and people invent new uses and mixes of modes, we seem to be at a loss: policy-makers and lawyers are forever running behind technologists, manufacturers and users, and the attempts to catch up with rapidly-changing communication processes are often awkward and even absurd. Thus, although it would in certain countries infringe copyright legislation, how would it be possible to prevent people from using widely available video-recorders to record television programmes off-air or off-cable for their own private use, except by police control which in many countries would be politically unacceptable and in many cases most probably unfeasible?

Similarly, if some ingenious individuals cobble together small satellite receivers with which to tap into satellite transmissions not intended for the general public, and soon afterwards manufacturers start to sell in the open market equipment for this very purpose, how should laws against such behaviour be enforced?

The second issue involves a set of profound problems in international law which so far has not been given due attention. Obviously, decisions by governments on national positions in international fora will be conditioned by factors such as politico-ideological attitudes and socio-economic conditions. However, a closer analysis of issues that are discussed in purely political terms reveals the impact of the varied, basic approaches embedded in different legal traditions. Thus, disagreements expressed in confrontational political terms may be shown to depend—sometimes to a surprising degree—on genuine but not articulated differences in legal approaches.

International law has developed from the basis of the Western legal tradition—but even within this tradition there are obvious differences between the civil law, the common law and the Nordic law systems. These differences concern not only the substantive provisions of law but also the approach to the law-making process. Thus, where the Anglo-American countries, following the common law approach, prefer to proceed pragmatically, formulating rules of legal behaviour as they acquire experience, the civil tradition tends to rely on the codification of rules in advance of action. This difference in approach to the making of law has been clearly demonstrated in the controversies surrounding the negotiation, in the United Nations Outer Space Committee, of international legal principles for satellite broadcasting. It is perfectly understandable that, apart from political differences of opinion, the common law-inspired reluctance to formulate rules in advance of actions can be—and has been—interpreted as politically-motivated prevarication or worse. The common law delegations would tend to regard the demands for early codification as being equally awkward.

Such differences in basic concepts and legal approaches

become even more pronounced when other, non-Western legal systems are also taken into account. Seen from a historical perspective, we face a situation where established branches of international 'communications law' such as telecommunications law and copyright law which first developed in the mid-1800s were formulated by a limited number of nations, mainly in Europe. Other applicable branches of international law such as space law have evolved in a very different and wider international context. Thus, at present and even more so in the future, a more pluralistic legal approach will be required: in an interdependent and multi-cultural world system, international law will have to broaden its approach so as to include philosophies and concepts from more sources than the Western legal tradition. Recently attention has focused on the international implications of Islamic law being adopted as the official legal system in such countries as Iran and Pakistan. There is a resurgence of interest in the various systems of African law and their relevance in such areas as self-reliant development and management of the environment. In this larger perspective there are striking differences in the very concepts of law and its role in conflict resolution between those countries that rely primarily on written law and court litigation for the settlement of disputes (for example, the United States) and those countries who have traditionally shown almost an aversion to written law and who rely on mediation, arbitration and other out-of-court methods to settle disputes (e.g. China).

International law generally, and specifically those branches applicable to communications and information, and to space, will therefore have to develop on the combined basis of three major considerations:

— first, international law must provide an adequate response to the implications of advances in science and technology;
— second, is the requirement for an evolution of international law from a multi-cultural perspective, and;
— third, is the requirement to provide an adequate response to the needs not only of countries, but of individuals and of the international community as a whole.

8 The great satellite game: round 3

'There's a new status symbol in the US; a radar dish aerial, 3 or 4 metres across, anchored on the owner's roof or in his backyard or garden.' (Fox, 1982, p. 680.) When aimed at a satellite used for the distribution of television programmes or other information and equipped with a lot of electronics, the antenna can reach more than a hundred satellite channels, absolutely free. Other viewers in the United States can also receive the programmes transmitted by satellite—but at a price. Normally the satellite-distributed programmes reach the viewer via local cable television stations which levy a monthly fee for a selection of programmes according to contracts concluded with the enterprises providing programme distribution via satellite.

The present boom in satellite distribution dates back no further than 1975 when a Pay TV service called Home Box Office provided the first satellite distribution to cable stations. The idea caught on, and after the usual wrangling between various interest groups the FCC, in 1980, adopted an 'open skies' policy. In principle, any one with some $70 million to spare can apply for permission to establish a satellite system for distribution of television programmes and other services. It can be risky—but there are now ten satellites in orbit that provide these distribution services and others such as entertainment programmes intended for reception by hotels and motels for their customers, teleconferencing and teletext. It is this seam of entertainment and information that the home satellite buffs are mining, through their 'television receive only', TVRO, equipment.

Since none of this material, including the television programmes transmitted for cable stations, is intended for the general public but for the intermediaries who ensure distribution to their customers, the owners of the TVRO stations are operating in a legal limbo. Some argue that it is theft to obtain access to programmes intended for others. Whatever the legal position, a whole new industry has grown up in America to service the requirements of this new sector of the audio-visual revolution. There are, it seems, already more than 50,000 TVRO stations in the United States and according to some estimates the number of such stations is increasing by about a thousand per month. This trend is no doubt helped by the rapid decrease in price, from some $40,000 to between $3,000 and $10,000 and rapidly approaching sums in the hundreds.

Satellite dishes and TVRO equipment are already on sale in Europe—but the situation in Europe is very different since television programmes are generally not distributed via satellite. So far, TVRO owners can pick up television programmes from the USSR and Saudi Arabia, neither of which are known for their entertainment value. However, Dutch cable stations have transmitted coverage of current events, otherwise not available in Europe. And a British entrepreneur has managed to get an agreement with Eutelsat, the satellite organization established by the European telecommunication administrations, to transmit programmes to cable systems in Malta, Finland and Norway. This is of course only a first step in a plan to get programmes to the heavily cabled countries in Europe, particularly Belgium and Holland. And there is still much confusion in Europe as to how to react to this new intrusion in the television landscape which, since the introduction of television has been dominated by national public service organizations. And this development is only one further complicating factor in the next stage: the arrival of operational direct broadcasting satellites (DBS).

Direct broadcast satellites (DBS)

Satellite broadcasting has always evoked ambivalent reactions, not least among the established broadcasters. No more than some 10 years ago one risked exile to the outer fringes of the science fiction ghetto if proven serious about satellite broadcasting. Surprisingly soon though DBS—as indeed science fiction—achieved respectability and even became somewhat of a bandwagon, brightly painted in white and black; white for the most euphoric of expectations, black for the darkest of fears. In the early discussions, the medium tones of reality were conspicuous by their absence. [Ploman, 1977, p. 82.]

The bandwagon, though, had its uses. Since it was highly visible, it attracted attention. Even if many of the reactions were shriller than they need have been, there was, for once, some thinking about possible uses and implications—prior to implementation. Not only technologists and hardware salesmen were abroad: international organizations, planners, diplomats, broadcasters and educationists found themselves involved, whether willingly or reluctantly. It also quickly became clear that engineers and technicians would be way ahead of prospective users and able to provide any required configuration—given enough incentive and money. Thus, the really interesting and important issues were not in the technological field; the prospective users and the regulators appeared bewildered and insisted on discussions of technical details which they could not understand when in fact the technicians were saying: tell us what you want to do and we will then provide the means and tell you how much they cost.

Despite the interest in DBS before it had become an operational fact, confusion has reigned supreme. Thus, when DBS issues were seriously taken up in the United Nations, in 1969, the United States and the USSR delegations first wanted to deny the feasibility of the technology and then the United States blandly declared that even if DBS were technically possible, there was no interest and no intention in the United States to use satellite broadcasting. When the FCC decided in 1982 to permit the establishment of domestic DBS systems in the United States, there were immediately some nine applications by different companies and more in the pipeline, probably so many that neither the available positions in the geo-stationary orbit nor available frequencies would be sufficient to accommodate them all.

This is a far cry from the position of 1969—much had happened in the meantime. In the 1970s the first experiments in the use of advanced satellites that could provide direct broadcast capability had been made: NASA's experimental satellite ATS-6 was used for the delivery of various services in the United States and India, and other countries such as Canada and Japan conducted their own experiments. The perceived need for internationally-agreed rules created much work and much controversy all through the 1970s —and will continue to do so.

A regime for DBS

The effort to work out an international regime for direct broadcast satellites is a particularly interesting example of the ways in which the international community actually copes with the challenge of a new technology. Which organizations are responsible? How do governments relate international policy and law to national policies and regulations? And how are the major issues defined?

Since responsibility for satellite communication matters is widely dispersed throughout the international system of organizations, issues were formulated in keeping with the particular objectives and approaches of each organization concerned. In the ITU issues were defined in technical-administrative terms and referred to such matters as allocation of frequencies and the associated orbital positions; avoidance of interference; procedures for notification and registration of satellite systems. In the United Nations the emphasis was on the further development of space law through the elaboration of legal principles in such areas as direct broadcasting and remote sensing via satelites. UNESCO debated issues concerning the uses of satellite communications while, at one stage, WIPO (World Intellectual Property Organization) considered the possible need for a special copyright regime applicable to DBS. Thus, in one context, issues concerned the technical regulation of DBS systems in the framework of general telecommunications law and, in another context, issues were defined in politico-juridical terms to which were added issues of content rules through which DBS issues were related to the debate on a new world information and communication order.

One of the difficulties facing the international community in evolving rules for satellite communications and in particular for satellite broadcasting has been the search for a minimum of coherence and consistency. It is a genuine difficulty. The fragmentation of responsibility and the differences of approach at the international level reflect similar diversity at the national level. Each international organization is linked to a particular national bureaucracy and its immediate 'constituency'. The technical regulations of the ITU reflect the traditions and conventions of national telecommunication authorities and are often not drafted in a manner which makes them easily applicable in other contexts. Conversely, the diplomats working through the United Nations might not appreciate the implications of technical developments. The problem is compounded when the same government might take one stand in the United Nations, another in the ITU and a still different one in other organizations.

Early ITU decisions

In the United Nations context, satellite communication was specifically mentioned in the early outer space resolutions adopted by the General Assembly. The provisions of the Outer Space Treaty of 1967, the bedrock of outer space law, are as applicable to satellite communication as to any other space activity. Simultaneously, in the ITU context, work started on the development of technical-administrative rules for space communications services. As early as 1963 an Extraordinary Administrative Radio Conference reached agreement on allocation of a limited number of radio frequencies for various space services. Technical work continued at an increasing rate so as to provide the basis on which decisions could be made for extending the range of frequencies allotted to space communications, for the allocation of specific frequency ranges to different space services and for the definition of the technical and other parameters of different services.

The ITU Radio Regulations in which are embodied the international rules regulating all forms of radio communications are the result of work that has been going on for more than a century. Each successive regional or world administrative radio conference adds further rules and regulations in keeping with its specific mandate. The Radio Regulations appear as an impressive building originally conceived in a late nineteenth-century European style which has been constantly enlarged by additions in new styles corresponding to the new technologies and services that required regulation. Each addition demanded consideration of the earlier parts representing the requirements of the older, established services and their claims on the radio frequency spectrum and on operational rules. The large, composite building representing the Radio Regulations in their current shape has become utterly confusing for the uninitiated who risk losing their way between principles of great general import and the technical-administrative rules laid down in the painstaking detail that is necessary to reach the sought-for consensus among the competing claims not only of countries but also of services. Whatever their disadvantages, the Radio Regulations have proved their worth and it would probably be almost impossible to reach agreement on a complete overhaul of the entire building.

A series of important decisions were adopted by the World Administrative Radio Conference for Space Telecommunications in 1971. This conference allocated radio frequencies for a wide range of space services and laid down new and complex procedures for the notification and use of frequencies and for the resolution of conflicts. As mentioned earlier, the ITU defined a number of different satellite communications services which by now have reached eighteen in number. At the time of the 1971 conference, technological and other factors made it seem reasonable and convenient to make a sharp distinction between the Broadcasting-Satellite Service (BSS) and the Fixed-Satellite Service (FSS) in which transmissions are intended for communication from one fixed point to another, between two defined correspondents, i.e. between two earth stations and thus between two defined end-users, be they television organizations or telephone subscribers. In contrast, the Broadcasting-Satellite Service was defined as a space radio communications service in which signals are 'intended for reception by the general public' (Radio Regulations, Art N1/1). This definition corresponds to the definition of terrestrial broadcasting whose transmissions are also said to be intended for direct reception by the general public. For broadcasting via satellites a specific provision was adopted in 1971 to indicate that direct satellite reception 'shall encompass both individual reception and community reception' (idem).

Individual reception is defined as 'the reception by simple domestic installations and in particular those possessing small antennas.' Community reception is defined as 'reception by receiving equipment which in some cases may be complex and have antennas larger than those used for individual reception, and intended for use (a) by a group of the general public at one location, or (b) through a distribution system covering a limited area' (idem). The first case may be represented by a school class, the second case by several classrooms in a school complex.

So far, so good. These definitions had only been adopted with considerable difficulty but the crux of the matter was another problem which had been given, according to some observers, unwarranted attention through action by the United Nations. In the late 1960s considerable anxiety was expressed by some countries over the prospect of satellite-borne television being transmitted from one country to another without any high-level regulation. There was also the desire to avoid a repetition of the admittedly chaotic situation prevailing in high-frequency, shortwave broadcasting: in the propaganda war on the air international regulation was hardly followed at all, or only to a limited degree, which caused constant problems of interference and put many countries at a disadvantage. In 1969, on the joint proposal of Canada and Sweden, the UN General Assembly decided to place the question of direct broadcast satellites on the agenda of the Outer Space Committee which to this end established a special Working Group on Direct Broadcast Satellites (DBS).

Work in the United Nations

The background to the United Nations action was very different from that of the ITU. In the ITU context rules for space communication services were seen as an extension of the regulatory activities of the Union in the field of terrestrial telecommunications, particularly with regard to the use of the radio frequency spectrum. The ITU approach was based on technical and administrative regulation where political—and to some extent legal—problems were played down or swept under the carpet. In the United Nations context space matters were first discussed in connection with issues of disarmament: the major concern was to prevent an extension of the Cold War and the arms race into outer space. Soon, United Nations involvement expanded and the General Assembly established a Committee on the Peaceful Uses of Outer Space (COPUOS). Originally, there were hopes that the United Nations would play a major role in promoting and even co-ordinating space activities at the international level but military, political and commercial factors favoured nationally-based space programmes, outside the United Nations. The main role of the United Nations and the Outer Space Committee came to focus on another crucial aspect: the development of the new branch of international law represented by space law. The major achievement in this area was the work to establish the bases for the Treaty on Principles Governing the Activities of States in the Exploration and Use of Outer Space, Including the Moon and Other Celestial Bodies of 1967 which, for obvious reasons, has become known as the Outer Space Treaty. This Treaty presents a number of unique features in international law: in this context it is important to note that the Treaty 'internationalizes' outer space by means of a prohibition against national appropriation in outer space. The work of the Outer Space Committee and its Legal Sub-Committee has since focused on the elaboration of agreements on specific aspects which are not covered in detail in the Outer Space Treaty, such as the rescue of astronauts and liability for damage caused by outer space activities.

This then was the background for the work on international legal principles for satellite broadcasting which began in 1969. From 1969 to 1974 this work was carried out by the Working Group on Direct Broadcast Satellites, which then felt it had pushed matters as far as it could in keeping with its mandate; this mandate could be seen as the task of mapping and clarifying the issue areas, identifying the different positions and generally providing the background for an elaboration of the legal texts which would be the work of the Legal Sub-Committee of COPUOS.

Which then were the main issues and why did the work proceed so slowly? There were many different sets of issues which were often interlocked and used in various combinations to promote stated or implicit national positions. In the first place, neither of the two superpowers was particularly keen on having the matter of satellite broadcasting discussed at all, and only yielded to pressure from the overwhelming majority of countries who insisted on the elaboration of some international legal rules beyond the technical-administrative rules that might be adopted within the context of the ITU. And then there were issues of procedure, i.e. differences of opinion as to how best to deal with this emerging new technology. The first major working papers prepared jointly by Canada and Sweden which provided the bases for much of the deliberations indicated the desirability and even the need to provide an international legal framework for an activity that might otherwise have disruptive effects and cause conflict in a politically sensitive area. This attitude was supported by the majority of the members of the Outer Space Committee and thus of the Working Group. In contrast, for reasons both of politics and legality, the United States, supported by some other countries—mainly of the common law tradition and in particular the United Kingdom—maintained that regulation should only be considered after the fact, i.e. after some experience had been gained in the field of satellite broadcasting and actual operations had shown the need for legal principles beyond technical agreements in the ITU context. Often, these complex attitudes were given a purely political interpretation. Those against regulation were seen as wanting to safeguard complete freedom of action, regardless of the wishes of other countries, particularly of those which would not be in a position to launch direct broadcast satellite systems themselves. Those who were in favour of UN-sponsored legal principles were seen as setting up obstacles to the development of a new and potentially promising technology and as wanting to introduce new principles in international law that might have unforeseen consequences in other areas, such as short-wave broadcasting.

Following the initial discussions on the probable development of the technology, which to a large extent have proved to be surprisingly accurate, and on a series of other matters such as the immediately applicable provisions of international law, one issue quickly emerged as the key one: the issue of what in shorthand has come to be called the 'question of prior consent' or 'prior agreement'. In summary, this issue concerned the question of whether one country should be permitted to establish a system and undertake television broadcasting via satellite directly to the public in another country, or other countries, without a prior arrangement with the 'receiving' country.

Some delegations talked about the need for rules governing the content of television transmissions via satellites; others preferred agreements of a more general kind, concerning the establishment and co-operation between the countries concerned; still others felt that there should be no rules beyond technical regulation. In any case, this issue was clearly related to the interpretation of the provisions concerning freedom of information laid down in the Universal Declaration of Human Rights and the Covenant on Civil and Political Rights. The unrestricted flow of information has of course been a key point in American foreign policy while other countries such as the USSR placed feedom of information below certain other principles of international law, such as non-interference in the internal affairs of other states. The matter is, however, more complex than is indicated by these extreme attitudes, and this is reflected in the fact that countries such as Canada and Sweden, with a traditional firm commitment to the principle of freedom of information, took the lead in the attempts to provide a legal framework for satellite broadcasting to other countries. In a somewhat summary form, it can be said that this case showed up certain contradictions in international law, or rather showed the need to find a balance between different but equally legitimate principles of international law which were set against one another: the principle of freedom of information and the free flow of information, as against the recognized right of each country to decide its own social and economic systems, including its systems of communication and information, and thus of broadcasting.

Direct broadcast satellites raised in a rather acute form new issues of sovereignty. Each country had evolved its own system of regulating broadcasting in terms of finance, rules of programming, complaints procedures, copyright rules and other aspects. Satellite television broadcasting direct to the public in another country might easily upset the often politically sensitive and sometimes rather precarious balance of internal interests which underlie the national broadcasting systems. Thus, for example, what would be the effect of the type of commercial television broadcast in the United States on countries which had decided on a totally non-commercial system without any advertising at all (e.g. Sweden) or which used carefully-regulated advertising (e.g. West Germany)? Even more serious were questions raised about political propaganda or content that otherwise might be unacceptable or shocking in a different cultural context. Obviously, the socialist countries which placed freedom of information at a different level in the scale of values than the 'liberal' West were adamant in their opposition to an unregulated flow of television programmes direct to their populations, which they tended to regard as an intolerable interference in their internal affairs.

Even more sensitive was the position of developing countries, most of which were without a complete national television network and where transmissions from outside might have major, unforeseen effects, politically, economically and socially. In this perspective the satellite broadcasting issue became linked to the debate on a new international information order in reaction against the uneven distribution of means of communication and the dominance of certain Western countries in the field of information. 'Information imperialism' or 'neocolonialism' were seen as associated with a continuing, even though different, form of political colonialism.

Back to the ITU

While the discussions in the United Nations continued, these politico-legal issues also spilled over into the ITU, despite all efforts to keep political questions out of the technical discussions. However, these issues could not be avoided even though they were discussed in a technical guise. At the 1971 Conference on Space Telecommunications a key issue was couched in terms of spill-over. In the ITU context it was assumed that satellite broadcasting would be undertaken on a national basis. However, a satellite beam or 'footprint' for national use cannot be made to fit exactly the borders of each country. The question was what to do about satellite-borne television signals that spilled over from one country into other neighbouring countries; however, the question was not discussed in terms of content and legal principles but in terms of the interference which might be caused to the telecommunications services in other countries. Whatever the terms of the dicussion, the issue was sensitive enough for agreement on all other questions before the Conference to be made conditional on the solution of this problem. In fact, a failure on this point might have jeopardized all other important results. After long and difficult negotiations, agreement was reached on what was termed 'technically unavoidable spill-over'. In a famous provision, which was inserted as item No. 428 A in the Radio Regulations, it is stated:

In devising the characteristics of a space station in the broadcasting-satellite service, all technical means available shall be used to reduce, to the maximum extent practicable, the radiation over the territory of other countries unless an agreement has been previously reached with such countries.

This rule has been interpreted by most countries and commentators as meaning that satellite broadcasting from one country to another requires the prior agreement of the recipient country and that, failing such agreement, only technically unavoidable spill-over would be permissible. However, the United States and some other countries have tried to draw a distinction between the technical nature of the regulations adopted within the ITU, and the more general aspects that are discussed in the United Nations. They took the position that 428 A was a mere technical regulation, confined to technical parameters.

Although technical developments tended to fudge the strict distinction between the Broadcasting-Satellite Service and the Fixed-Satellite Service, the ITU continued to use this technical distinction as a basis for its regulatory work. Also, its work continued on the assumption that satellite broadcasting would be for national purposes only. In 1977, at a World Administrative Radio Conference on Broadcasting-Satellites, a plan was adopted for direct broadcast satellite systems in respect of geo-stationary orbit positions and the use of the most important frequency band allocated to this service (around 12 GHz). This plan was adopted for all regions of the world except the Americas, for which a similar plan was to be negotiated at a later stage (1983–84).

The 1977 plan allotted a minimum of five satellite TV channels to almost every country, as from January 1979. It laid down details of the coverage areas, the positions of the geo-stationary orbit and other technical characteristics of each allotment. It is these 1977 allotments that form the basis in international law of the current plans in Europe and elsewhere for direct broadcast satellite services. The legal basis is thus the ITU sponsored technical regulations which, however, have implications beyond the purely technical-administrative context.

The results of the Conference imply that each country, by approving the 1977 plan, has indicated its agreement to the specific coverage areas assigned to each country, both its own and that of other countries, including the technically unavoidable spill-over. The plan, in principle, provides only for national satellite broadcasting. International television broadcasting via satellites was requested and approved only in certain specified cases: for the Nordic countries of Denmark, Finland, Iceland, Norway and Sweden for their projected Nordsat system; for Tunisia, Saudi Arabia and Syria for a possible Islamic network; and for the Vatican State for coverage of all Italy.

And back to the United Nations

In the meantime, discussions continued in the United Nations. In 1972, the General Assembly, following a USSR proposal for an international convention on satellite broadcasting,

reaffirmed the desire for international legal rules in a resolution adopted by an overwhelming majority (102 votes in
favour, 1 against (USA) and a few abstentions). However,
neither the Working Group nor the Legal Sub-Committee
managed to reach the necessary consensus on the key issues
of 'prior agreeement'.

The failure on this point was vividly described in an
article by two Canadian experts, John Chapman and Gabriel
Warren, then senior officials in the Department of Communications. They start their brisk and sharp analysis in a
tone of exasperation:

> In what has unfortunately become an annual occurrence, the 1979
> sessions of the Outer Space Committee and its Legal Sub-Committee
> have again come and gone without a consensus having been reached
> on the draft principles on Direct Broadcast Satellites. Again there was
> a sterile debate on the major outstanding issue, i.e. the balance that
> should be struck between the 'free flow of information' and the right
> of the receiving state to agree before television signals can be beamed
> from the satellite of one state directly into individual homes or small
> community receivers in the territory of other states . . . As each year
> passes without the deadlock being broken, the Outer Space Committee
> loses credibility as the focal point in the UN system for taking an over-
> view of all aspects of outer space issues. [Chapman and Warren, 1979.]

The Legal Sub-Committee has thus, despite all efforts, been
unable to achieve the required consensus on the basis of
several draft sets of principles. As could be expected, the
Soviet and the American drafts represent opposite extremes.
The Canadian/Swedish drafts tried to bridge the gap and
provide a text which would correspond to the wishes of
the majority of members in the Sub-Committee while safeguarding the interests of all. In the opinion of the two authors
mentioned above, the USSR had, at the 1979 session
'bent over backwards' to assist in breaking the impasse.
They therefore address themselves mainly to the United States
of America:

> If, as it now seems clear, DBS will not be used for transmitting pro-
> grammes in the same unregulated way as international shortwave
> broadcasting in the High Frequency (HF) band, why does the USA
> continue to oppose any UN principle embodying the right of the receiv-
> ing state to agree? . . . The USA seems to view the DBS issue as one

aspect of a growing pattern of restraints being promoted under the umbrella of 'a New International Information and Communication Order' . . . Why can't the US agree upon a set of principles compatible with the 1977 ITU plan which will not harm the essential interests of any country or group of countries? . . . With a number of countries on the verge of introducing operational DBS systems, it would be reassuring to know they have the blessing of the UN and not merely the ITU. [Chapman and Warren, idem.]

The matter became even more complicated through what appeared to be unco-ordinated and confusing behaviour on the part of certain countries—a fact which is generally not referred to in polite diplomatic circles. It is instructive to analyse the positions that some countries have taken on the issue of satellite broadcasting in different international organizations. While some countries have been consistent all the way through (e.g. the USSR), it is not difficult to find examples of a country taking one position in the United Nations and a quite different one in the ITU (e.g. West Germany). This reflects the difficulties that countries face in solving differences of opinion at home or in trying to co-ordinate the views of different government departments and their constituencies. Thus, in the United Nations some countries would oppose the principle of 'prior agreement' in the name of freedom of information. And in the ITU the same countries would oppose the idea of international satellite broadcasting as being detrimental to their national broadcasting systems and, instead, approve in technical terms 'prior agreement' as expressed in item 428 A of the Radio Regulations and in the 1977 plan. Some delegations in one forum were therefore unaware—or pretended to be unaware—of what happened in another.

There is another recent development which should be a lesson to those who oppose 'prior agreement'. There is only one case where a group of countries have studied in depth and in real terms the kind of agreements required for orderly satellite broadcasting, at the international level; that is the preparatory work for a possible Nordic system. Legal studies have shown that the arrangements that would be required in real life are wide-ranging and complex, and cover a range of issues from copyright via civil and criminal liability to complaints procedures over programme content. The Nordic countries have not found this work at all easy, and yet they are five countries which are closer than most in shared legal approaches and a long tradition of close co-operation and co-ordination in legal matters.

Current results

It is against this background that should be seen the outcome of the drawn-out work in the UN Outer Space Committee on legal principles for satellite broadcasting. Despite all the impatience shown by delegations over the apparent impasse on the issue of prior agreement, there was at the same time a will not to upset the 'gentlemen's agreement' following which the Outer Space Committee works according to the principle of consensus and not according to vote. This practice goes back to the early times of the Committee's existence and represents a compromise agreed upon so as to ensure the participation of the two major space powers in the work of the Committee; obviously, international agreements concerning outer space to which either of the superpowers would not accede, would have very limited usefulness.

However, in 1982 this understanding was broken. At its meeting in 1982, the Outer Space Committee was faced not only with the continuing lack of agreement on substance, i.e. on the principle of 'prior agreement', but also with controversy in respect of further negotiations: some states felt that the principles so far evolved should be adopted by the General Assembly, if necessary by a vote, while other states favoured further attempts at achieving consensus. Acting against the established practice of only introducing resolutions in the General Assembly when all members of the Outer Space Committee had agreed, a text was introduced by a number of countries under the title 'Preparation of an international convention on principles governing the use by states of artificial earth satellites for direct television broadcasting'. The effect of this move was that delegations in the General Assembly in fact voted on two matters, i.e. on the overt question of substance and on the underlying question of the consensus practice. The result was that only eighty-eight delegations voted in favour of the resolution, fifteen against with eleven abstentions. Among the latter groups were also to be found not only countries which had consistently been against the adoption of these principles, but also countries which, with equal consistency, had worked for the adoption of international legal principles on satellite broadcasting.

The principles adopted by the General Assembly contain sections on purposes and objectives, the applicability of international law, rights and benefits, international co-operation, peaceful settlement of disputes, State responsibility, duty and right to consult, copyright and neighbouring rights, notification to the United Nations, and consultations and agreement between states. It is this last section which covers the contested principle of 'prior agreement'. As adopted, the text reads:

1. A State which intends to establish or authorize the establishment of an international direct television broadcasting satellite service shall without delay notify the proposed receiving State or States of such intention and shall promptly enter into consultations with any of those States which so requests.

2. An international direct television broadcasting satellite service shall only be established after the conditions set forth in paragraph 1 above have been met and on the basis of agreements and/or arrangements in conformity with the relevant instruments of the International Telecommunication Union and in accordance with these principles.

3. With respect to the unavoidable overspill of the radiation of the satellite signal, the relevant instruments of the International Telecommunication Union shall be exclusively applicable.

If victory there is in having forced through a vote, it is something of a pyrrhic victory which might have negative effects on the future work of the Outer Space Committee.

The ITU has also continued its work: in the summer of 1983, a Regional Administrative Conference was convened to adopt a plan for the Broadcasting-Satellite Service in the region of the Americas. Despite the controversies which opposed, not as could be expected the North and South but rather Canada and the United States of America, it proved possible to achieve agreement. Advances in technology have made it possible to adopt technical bases for this regional plan different from those accepted for the 1977 plan for the rest of the world, which is now described by some analysts as outdated and backward. The 1983 regional plan shows more technical flexibility and accommodates a great number of national channels spread over various time zones.

Both in the United Nations and in the ITU discussions have continued on the status of the geo-stationary orbit, following demands by the equatorial states to exercise ownership or at least control over the orbit. In scientific terms, these demands do not make sense since the positioning in the geo-stationary orbit depends on an equilibrium of gravitational and centrifugal forces of the planet. They are now based on the concept of the orbit as a limited natural resource which the states concerned should control as any other natural resource. This position is strenuously resisted by most other countries as going against one of the basic principles of the Outer Space Treaty prohibiting national appropriation of outer space. Furthermore, the regulation of the use of the geo-stationary orbit is inexorably linked to the use of radio frequencies: regulation for one requires the interlinked regulation of the other. The next round in this negotiation will therefore occur at the overall Space Communications Conference which the ITU will convene in two sessions, 1985 and 1987, when all outstanding and disputed issues will be taken up, probably including new ones that have arisen in the meantime.

DBS for real

For years it seemed that action on DBS was confined to numerous studies and proposals, to the experiments carried out by a few countries and to the regulatory decisions in the ITU and the drawn-out debate in the United Nations. Then, recently, it was as if a dam had burst, letting loose a torrent of plans and decisions to establish DBS systems. The current is still unruly enough for plans to emerge and disappear in rapid succession, even more so since the DBS current encountered other currents and cross-currents in the shifting river waters of new technologies and media. It seemed as if, suddenly, the industrialized countries had become mesmerized by the communications and information sector, as the saviour from economic stagnation, as the focal area for industrial innovation and employment opportunities. If for some years the primary emphasis had been on telecommunications, computers and informatics, the turn had now come for the audio-visual sector to emerge as the latest additional stream in the mighty river which, inexorably it seemed, carried all to the promised goal of the information society.

While individually, countries often, in what appear to be moods of almost desperate euphoria or concern, try to sort out their policies and plans of action, national frontiers become ever more transparent. The margins for independent, national decision-making become more limited and also more complex when national decisions impinge on each other, often in unexpected ways. Only a few of the largest and most powerful countries seem relatively impervious to outside influence even though action in the international political arena and the international commercial market place requires some attention to what other countries are doing.

In the United States, the key concepts seem to be deregulation to let loose competition, between technologies, services and entrepreneurs, as the prime mover towards the brave new information world. To be sure, the official reasons for introducing DBS in the United States mention various services which would be of benefit to the public, such as increased television service to remote areas, additional television channels throughout the country and innovative services such as high definition television. Not all observers are convinced: most of the planned DBS systems do not seem geared to provide anything but more of the same: more of the same fare now provided over terrestrial broadcasting or over cable systems. The exception seems to be the plans of a system which is proposed by a company appropriately entitled the Direct Broadcast Satellite Corporation: very high-powered satellites with very low-cost receiving equipment would be operated in such a way that producers and providers of programmes could get access for as little as $6,000 an hour for nation-wide coverage, and a mere $200 an hour for covering the Boston–Washington corridor. Such a system, if ever realized, would change the rules of the television game in the United States by permitting access to others than the big corporations. Thus, while this system 'offers a maximum access to a nationwide network at lowest cost, DBS systems proposed by other companies are just another example of further concentration of media interests' (Brewin, 1982, p. 22). In any case, the needs or wishes of the individual citizens who make up 'the general public' are not mentioned, other than as objects of market analysis, whether from a commercial or a public service perspective.

Canada seems intent on creating or maintaining an audio-visual space of its own, withstanding the pressures from the colossus to the south. Japan studied and adopted long-term policies—and acted upon them: integrated information network systems, policies for the diversification of broad-casting, and for the fifth generation computers, robotics and artificial intelligence, grand visions of man, communications and computers in the future information society.

The hopes and fears in Europe are different: geography, cultural diversity and history give a different perspective. The situation in Western Europe is, above all, confused—and confusing, nationally as well as regionally. France, worried about the prospects for commercial DBS predicted that European cultural values would become an extinct species unless the European governments would agree to a vaguely defined 'espace européen audio-visuel'; at least the concept seems 'excellent in stressing that from now on all European television stations are visible and have to co-exist together in one limited area, like goldfish in an aquarium'. (Hondius, 1983, p. 159.)

What is the problem exactly, since the 1977 agreement within the ITU provides for national coverage and a prohibi-tion of radiation over other countries beyond what is techni-cally unavoidable? The reality is of course more complex than these rules. The 1977 ITU plan outlines strict technical parameters to ensure that a country's national broadcasting service can be received within that country by any 'standard' satellite receiver—and the standard was agreed to be a 90 cm. diameter dish. Taking this focused service area as the base, the satellite power density falls off progressively with increasing distance from the focus of the coverage area. On the basis of all technical parameters it is possible to compute the 'footprint' of each satellite beam for reception by the standard antennas—but also by antennas of a greater size. Thus, the footprint of the Luxembourg satellite would reach at least some 30 million television homes with the use of a one metre dish. The calculation for an Austrian satellite has shown that using a reception antenna of 60 cm. the Austrian satellite signal could reach some 44.8 million people living in Austria and surrounding areas in West Germany, Switzerland and Italy; with an antenna of 90 cm. the signal would be receivable by some 111 million people, with the addition of reception areas in Czechoslovakia, East Germany, Hungary, Poland, France and Yugoslavia. Conversely, with the use of 90 cm. dish antennas, the public in Austria could receive signals from West Germany, France, Switzerland and Luxembourg. Similarly, the French, West German, British, and Swiss satellite signals could, to varying degrees, be received in Spain.

This creates a totally new situation in Europe. So far, technical characteristics of terrestrial broadcasting have favoured and still favour certain kinds of national political, financial and managerial responses, as well as certain kinds of cultural and aesthetic responses. Development has been from local to national, with the addition of some transborder television through overspill and cable distribution, but only in contiguous countries. With DBS, the very ground of national cultural and communications independence seems to shake.

However, it is not only DBS which is creating media turmoil in Europe. While some countries, such as Belgium and the Netherlands, already have extensive cable television networks, a new rush to wire for cable television is on: governments anxious to create new industrial and employment opportunities and to establish national electronic information grids are pushing plans for the cabling of their countries, with Britain and West Germany in the lead. On top of this, there is the need to grapple with the impact of home video and telematics, and with changes in the telecommunications structure brought about by new technology and pressure from the United States for de-regulation also in other countries. None of the European countries can act in isolation: decisions by other countries affect the margins for domestic freedom of action. The attempts by regional organizations such as the EEC, the Council of Europe and the OECD to achieve a measure of coherence and commonality of policy seem largely ineffective. No two countries seem to have the same outlook or priorities, nor do the different interests involved, be they entrepreneurs like the press magnate Rupert Murdoch entering the satellite television business on both sides of the Atlantic, national broadcasting organizations, copyright owners or local governments acting as cable operators. Industry prevails over culture and communications: little thought is given to the content to be transmitted over all these new channels and outlets. Typically, while the first experiments in European, multi-lingual, pan-national television broadcasting are launched (Eurikon), national positions seem to preclude a common technical standard for colour television broadcasting via satellites.

Thus, the United Kingdom tries to cope with the implications of its telecommunications de-regulation policy, and with a combination of industrial and media policy for cable television and DBS but succeeds no better than having its satellite television being called 'a melodrama looking for a script in the jungle of the media revolution' (*The Guardian*, 19 September 1983). In 1981 the BBC was awarded Britain's first two channels for satellite broadcasting over a British-built high-powered satellite at an annual rental of £24 million. After much agonizing, the BBC in December 1983 came to the conclusion that its plans to begin DBS services in 1986 were hopelessly uneconomic. And in the meantime, the authorities worry over what the Americans call 'backdoor DBS', i.e., home reception from telecommunications satellites, which in 1984 will be offered over Mr Murdoch's Skychannel service.

In continental Europe a major DBS story has been the distress—which to some observers amounts to paranoia—with which France and Germany have reacted to the Luxembourg satellite plans, which are seen as detrimental to the national television services. Although France has found her ambitious plans in telecommunications and informatics more expensive than expected and might even face a delay in satellite deployment, the government has given priority to the development of the communications industries to the tune of Frs.21,000 million: promotion of creativity in television and cinema, production of computer software and videogames, cable experiments, professional training in relevant fields and support for a determined export drive.

The complications of the satellite game are also shown by analysis of the negotiations for the stalled Nordsat project. It became clear that this project had three political dimensions: the politics of culture, the politics of communications and the politics of industry. Each of these systems has its own interests and its own rules of the game, which makes co-operation difficult.

Since public broadcasting traditionally belongs to the spheres of both culture and communications, their forms of concurrence . . . [are] rather involved. Between the communications sector and the industrial sector there ought to have been some basis for dialogue since their interests in the Nordsat project coincided in matters of technique. Nevertheless, there was a strong sense of rivalry between them . . . Between cultural interests and industrial interests there yawned an abyss. There seemed to be no way of talking together, of agreeing jointly and taking decisions. How can one weigh values of an increased understanding of language and 'cultural co-ordination' against market shares and employment opportunities? [Lindgren, 1981, p. 58.]

The stakes seem to rise increasingly higher. Western European countries, individually and collectively, have not seemed able to cope with the implications of the American policy of de-regulation in the communications and information field and the expressed desire to press other countries into similar de-regulated, commercially competitive patterns. Not only Europe but also other countries such as Japan are subject to this pressure which, if successful, is seen opening new markets for hardware and software. The responses of European governments seem inadequate and confused in a situation where the real issue, in the view of some observers, concerns the survival of the public service model in communications.

The new wave is also hitting other countries. India, for

example, is embarking upon an ambitious programme for the extension of television at the level of some $400 million over the next few years, but the bulk of the funds seems to be for hardware. In this and other countries, the large and often illegal import of video equipment again shows up the inadequacy of current policy and practice, reflected in such responses as an increase in airtime given to the commercial Hindi cinema on Indian television. It is significant that in Canada questions are now raised about the implications of the public policy focus on developing means of communication, on new channels and outlets which are then being filled with American programmes. Thoughtful observers everywhere criticize the inadequate consideration of sociocultural and long-term consequences of a headlong rush into a new information society dominated by communications hardware.

9 Satellites and the development of international relations

Admittedly, the picture is confused. As the number of players increases, their inter-relationships grow more complex. How can we make sense of the confusing satellite game as it affects countries and their relationships; what intellectual tools do we have at our disposal to disentangle at least some of the strands in these processes? If we look for help on these questions to communications theory and research, we will be strangely disappointed: most efforts seem to concern specific aspects of the international flow of information, often linked to the controversies about a new world information order. Instead, the focus will be on recent ideas that have been developed in international relations study but that have generally not been applied to communications or to satellites. We can begin by trying to discern some of the general trends that are said to characterize the current evolution of international relations and see how well—or how badly—they apply to satellite communications.

Traditionally, international relations analysis had almost exclusively focused on the activities of states in terms of power balance, military security and other elements of 'high' international politics. In recent times, international affairs theory has had to widen its approach in order to cope both with the changing international system and with emerging new ideas, new ways of looking at the international scene. Economic policy issues and international economic relations are now seen to condition 'traditional' politics. Environmental issues, problems of resources, population and energy have been placed high on the international agenda. During the last few years the development of science and technology and their implications have created new problems and opportunities. Satellite communications, which are located at the intersection of the new communications and information technology and outer space activities, represent a prime example of these new kinds of problems facing the international community.

A major impact of science and technology can be seen in the increased politicization of issues. This is understood to mean that a growing number of issues and activities are placed within the area of conscious social choice and are thus subject to political policy-making. There are many reasons for this trend. With the increasing power of science and technology more and more areas are brought under human control or become subject to human intervention. Phenomena and events which were earlier ascribed to acts of God or to the vagaries of nature are now seen to be affected by human activities (e.g. pollution) or to result from conscious human decisions (overfishing). There is thus a growing international concern over the control and management of resource realms that have been considered beyond national jurisdiction (larger parts of the oceans and sea-beds and outer space); that advancing technology can now affect (weather and climate) and that otherwise present features beyond the conventional physical resource concept (radio frequencies, geo-stationary orbit, information).

One effect of these developments is that decisions in areas which previously could be or would have been left to individual, i.e. private, initiative are now moved into the sphere of public, i.e. political decision-making. It would obviously be impossible to leave to individual decision activities that might adversely affect the weather or cause chaos in the use of radio frequencies or risk collision between satellites. Thus, there is a need to extend public decision-making which in effect means to extend the powers of the state as the only public decision-making machinery available to us. As can be seen from the resource areas mentioned above, this trend towards increasing politicization of issues impinges on satellite communications from all sides. Even though in countries such as the United States there is a trend towards increased private commercial activity in outer space, various factors such as international commitments or military interests will still require public, i.e. government, decisions for policy framework, regulation and control.

Radio frequencies and the geo-stationary orbit are both resources, in the sense that without agreed rules of use they cannot be effectively utilized by anybody. Without rules for the allocation and use of radio frequencies the various users would interfere with each other and render this valuable resource useless.

The same is true of the geo-stationary orbit. In a very real sense these resource realms cannot be made subject to physical appropriation: there can only be a regulated, ordered right of use or no use at all. Also, since these resources are not unlimited, they will have to be shared and allocated to different users—and such allocation has always been a major function of political decision-making.

Thus, this concept of a growing politicization of issues is applicable in the case of satellite communications. The increasing demands on the radio frequency spectrum and the geo-stationary orbit can only strengthen this trend. The increasing militarization of outer space has added a further political dimension to outer space activities generally, as has, in a positive manner, the need for international co-operation in meteorological and environmental sensing of the earth via satellites. It is therefore a safe prediction to foresee an increased rather than a decreased politicization of issues relating to satellite communications. If this holds true from the space perspective it is equally valid from the perspective of communications and information, which in recent years have also been placed high on the international political agenda, as reflected in the new order debate.

There is another process that will reinforce this trend: the increasing linkage between international and national policy, and politics. According to the classic, so-called 'realist' model of international relations, states were perceived as closed, impermeable and sovereign units, each unit being roughly equivalent to all other units. In this 'billiard-ball' model of the inter-state system a strict distinction was made between politics within nations and politics between states. This distinction was expressed in the perceived differences between intra-state and international behaviour: within countries—a legal order, a rule of law; between states— the law of the jungle.

This 'realist' model has been challenged on all essential points. Sovereignty is no longer accepted as absolute and indivisible but as a relative concept that is used to encompass a great variety of national 'actors' from the superpowers to mini-states which differ markedly in size, resources and capacity to act independently. Thus, in general terms, nuclear power, space power and information power are concentrated in the same states. Hence the risk of new patterns of dominance and dependence and the extension of inter-state relations perceived in terms of centres and peripheries.

Equally revealing is that the strict separation between national and international politics is no longer true, if it ever was. In fact, the linkage between national and international policy- and decision-making seems to be the rule rather than the exception in such new areas of concern as the environment, outer space, communications and information, and therefore by definition in satellite communications. This interlinkage has been strikingly described as the growing internationalization of domestic issues and the domesticization of international issues. The see-sawing between different national interests combining in various alliances at the international level has already been mentioned. The result may be, as developments in Europe show, great difficulties, sometimes paralysis in the capacity to achieve common objectives at the regional level. Thus, international issues cannot be solved without constant reference to the various domestic interests. Conversely, domestic issues cannot be solved without constant attention to international developments that constrain or otherwise influence the margin of national independent decision-making.

These phenomena can be added to those mentioned earlier: the growing number and variety of transnational actors, the sectoral approach both to outer space and to communications and the overemphasis on technology when dealing with new information services. In looking at how national societies and the international community have responded to the new challenges posed by communication satellites, it was easy to discern the reliance on single points of view when what is needed are multiple points of view. Similar to what is happening in many other areas, the basic question seems to be: how we can learn to manage new levels of complexity which at present seem beyond our grasp?

A new world information and communications order

The difficulties we encounter in dealing with new and complex issue areas at the international and at the national level for that matter, are well exemplified by the debates on the new international information order which have also concerned satellite communications, and in particular satellite broadcasting and remote-sensing via satellites.

An initial problem is that the concept of a new international information order is used in two different and often unrelated ways. In one sense the concept is used to denote the changes that are brought about by the informatics revolution, the impact on individuals, societies and relations between countries of the increasing use of computerized data systems, the spread of microprocessors and the link-up between computerized data banks and telecommunications.

In another usage, the expression has been given a wider and looser sense as reflected in the debates in UNESCO and other international fora. Unfortunately, this debate has become as confused as it is controversial. Initially, the focus was on the flow of news and to some extent on entertainment materials. The developing countries, supported by the Eastern European nations, accused the Western industrialized countries of abuse of such concepts as freedom of information and free flow of information to further their own interests and to maintain their dominance over international information flows. Thus, one aspect concerned the biased and unfair reporting to which the developing countries felt themselves subject; another referred to the uneven distribution of communications media in the world. Often the attempts made by developing countries to correct the obvious imbalances in the international distribution of communications resources and information flows have been bewailed by Western media and some governments as attempts to extend censorship and government control over the media. While some positions quickly hardened or even petrified, there were also signs of a dawning recognition that issues were more complex than originally envisaged and that more sophisticated approaches were required. However, the new order concept soon came to attract and to represent an aggregate of varied issues, phenomena and factors which were often indiscriminately thrown into a boiling pot grandly marked 'the new world information and communications order'.

There are further problems. It is revealing that the debates on the new international information order—just as the debates on national communications and information policies —have been conducted with a great deal of heat but without any clarity as to what is to be covered by the word 'information'. It is therefore difficult to get any grip on 'order', whether old or new. In dealing with 'information' as an undefined object which mysteriously exists 'out there', no attention is paid to information, not only as an object or product but also a process and activity, nor is any distinction made between different kinds of information flows. To begin with, much of the controversy focused on a rather narrowly perceived, 'journalistic' dimension of news and to some extent on entertainment: the concern was with what might be called cultural sovereignty and identity. The views held on this particular dimension were then applied to other information flows without any rigorous analysis of the specificity in function and objectives.

However, it is quite obvious that countries do not deal in the same manner with information flows of different kinds: the dissemination of health or disaster warnings would be dealt with differently from entertainment films. The same is true at the international level. This means that we are in fact dealing not with one order but with a series of orders conditioned by the purpose and content of the information flow. Thus, for certain information flows there already exist international information orders which are not only accepted but actively supported by all the countries involved. Examples are the epidemic warning system set up by the World Health Organization and the meteorological information systems established by the World Meteorological Organization which have been mentioned earlier.

One discernible strand in the debate thus concerns such

factors as information dominance, national sovereignty and cultural identity. However, even on this point the issues are more complex than a simple North–South confrontation, as shown by the debate in the United Nations on legal principles to govern satellite broadcasting. It was in fact two industrialized countries, Canada and Sweden, who took the lead in the drive for an international legal regime that could provide a balance between the principles of free flow of information and the right of each country to decide its own communications and information policies.

The issues of 'cultural domination' in combination with the unequal distribution of communications infrastructure in the world resulted in the extension of the concept of a new international economic order to the field of information. As stated by the Fifth Summit Conference of Non-Aligned Countries in 1976: 'A new international order in the fields of information and mass communications is as vital as a new international economic order.' The growing recognition of existing imbalances between different parts of the world and within each one of them has led to decisions for joint, practical action to assist the countries that are most deprived as regards modern means of communications, as for example through the ITU technical assistance and the International Programme for the Development of Communication established within the framework of UNESCO, in 1980.

In the meantime new dimensions had been added to the international information debate. The impact of new communications and information technology on human rights was expressed in the need to extend the protection of individual privacy, particularly in the case of computerized data systems and transborder data flows. In this area, European countries took the lead and concluded in 1980 a convention under the auspices of the Council of Europe. However, new issue areas seem to arise at a speed equal to that of the spread of technical gadgets. Already the emphasis is shifting: a new focus of international attention and controversy concerns the politics and economics of the international flow of financial, economic information and data—which to a large extent, and increasingly so, are transmitted via satellites. The storm signals are already up:

there are warnings about new types of trade wars, new confrontations—mainly opposing the multinational companies and their countries of origin, i.e. mainly the United States of America, which advocate the largest possible freedom of action; and all those countries that are intent on balancing this 'freedom' with national economic, financial and political concerns.

Here we touch the issues which are crucial at a different level from that of the debate on freedom of the press. Two examples will show the gravity and the complexity of the issues we face—typically, both examples are linked to the use of satellites. The first example concerns the use of satellites for remote sensing. In the UN Outer Space Committee debates have focused on the rules that should govern not the collection of data via satellites but rather the use of the information gained from these data. Should the American space agency NASA have the right to sell information about another country—information that might reveal oil or mineral deposits—to a third party, say an American mining or oil company, without the knowledge and approval of the country in question? Similarly, the information gained from satellite data about the failure of the crops in certain parts of the world will make it possible for a grain dealer to buy in the commodity exchanges in the expectation of higher prices from needy countries. The second example is equally disquieting. The data networks established by commercial banks make possible not only the transmission of information but also instantaneous transaction of financial deals —at a speed which totally circumvents the possibilities of action by the national financial authorities, be they ministers of finance or central banks. As stated even by bankers, the banks find themselves suddenly facing situations which admittedly are of their own creation but which, because of the speed and volume of the information flows, have an unexpected and often uncontrollable impact.

Some of these issues also make it clear that the influence of communications technology, particularly in combination with computer technology, has become a principal agent in transforming modern conceptions of the nature of sovereignty. The impact of these technologies has been a decisive factor in transforming traditional views of sovereignty which have been understood and expressed in geographical and territorial or spatial terms into a new kind of concern that has been defined as a concern about communications or informational sovereignty or integrity.

In examining the policy and legal implications of direct satellite broadcasting, remote-sensing via satellites or transnational data transmission and storage via interlinked computers, we see another instance of the process by which an attempt is made to balance the fundamental but often opposing or competing values of sovereignty on the one hand and the flow of information and ideas on the other. To approach this issue, it is, first, necessary to dispel the notion, almost the mythology, that there is any state, however 'libertarian' its policy, that does not control information through rules on such matters as obscenity, public morality and order, sedition and national security or protection of individual rights. Similarly, every state has an international commitment to regulate from a technical point of view its radio communications, including broadcasting. In fact, the two principles and concerns co-exist in the preamble to the International Telecommunication Convention where the contracting parties agree to the object of 'facilitating relations and co-operation between the peoples by means of efficient telecommunications services' while at the same time 'fully recognizing the sovereign rights of each country to regulate its telecommunications'. Therefore, it seems necessary to seek for a fair blend among the interests of the individual in safeguarding his rights, the interests of the state in protecting its sovereign rights, and the interests of the international flow of information.

Setting satellite communications in this wider context we encounter a triad of key issues that underlie current perceptions and developments: vulnerability, security and interdependence. Each of these concepts represents a shorthand for a complex set of interlocking issues and each is used to cover a diversity of meanings. They are all interrelated and indicate phenomena and trends that are crucial for individual societies and the world at large.

Vulnerability

The efforts to achieve new forms of security and the new perceptions of interdependence are closely linked to emerging notions of vulnerability or—better expressed—to new dimensions of vulnerability which have recently come to be perceived at the national and international level. New forms of vulnerability are often seen as inherent in the conditions of modern society. Some kinds are obvious and have been dramatically demonstrated: a minor technical error led to an electricity black-out which affected a large part of the north-eastern United States; the chemical explosion at Seveso in Italy caused the evacuation of an entire region, and the accident at Three Mile Island nuclear power station caused a scare of a new kind. Terrorists can hold an entire society to ransom, and industrial action in a key sector may disrupt an entire economy. Another kind of vulnerability has been symbolized in the figure of Dr Strangelove, a madman in a position to unleash the ultimate catastrophe—a nuclear war.

It is obvious that the increasing militarization of outer space has caused a new sense of vulnerability. This is clearly expressed in the resolution adopted, after intense negotiation, by UNISPACE 82: 'The extension of an arms race into outer space is a matter of grave concern to the international community. It is detrimental to humanity as a whole and should be prevented . . .' As mentioned previously, communications and surveillance satellites play a key role in this regard.

Interestingly enough, new dimensions of vulnerability have to a large extent been perceived in relation to developments in communications and information. The reasons for a sense of vulnerability among developing countries faced with the overwhelming dominance in hardware and software by some industrialized countries are obvious and easily understood— even though many in the West refuse to marshall the required understanding and empathy. Perhaps even more insidious and difficult to deal with is the sense of 'communications vulnerability' among some of the technically most advanced countries.

This sense of vulnerability has led a country such as Canada to establish a 'Consultative Committee on the Implications of Telecommunications for Canadian Sovereignty', which became known—after its chairman—as the Clyne Committee. The thrust of the Clyne Committee's approach can be illustrated by some quotations from its report published in 1979:

Canadian sovereignty in the next generation will depend heavily on telecommunications. If we wish to have an independent culture, then we will have to continue to express it through radio and television. If we wish to control our economy then we will require a sophisticated telecommunications sector developed and owned in Canada to meet specific Canadian requirements. To maintain our Canadian identity and independence we must ensure an adequate measure of control over data banks, transborder data flow, and the content of information services available in Canada. If we wish to build a Canadian presence in world industrial markets, then we will be required to encourage the growth of Canadian telecommunications industries that will be competitive in world terms . . . We see communications as one of the fundamental elements of sovereignty, and we are speaking of the sovereignty of the people of a country. [See Consultative Committee report, 1979.]

It could not be better put by any representative of a developing country: just change the name of the country and it is exactly the same reasoning which many Westerners find so unpalatable coming from Third World countries.

The same reversal of position has happened over another issue in communications. In the discussions on legal principles for satellite broadcasting conducted in the UN Outer Space Committee, countries such as the United Kingdom and West Germany have shown, at best, lukewarm understanding of the demands that television broadcasting via satellites from one country to another should require prior arrangements between the countries concerned. Not only developing countries but also countries like Canada, France and Sweden have actively pursued this policy, which in the name of free flow of information has been resisted primarily by the United States. However, when it became known that France and Germany would establish national broadcasting satellite systems and that Luxembourg was considering a satellite service of its own, all of which might be received in the United Kingdom, some reactions in the House of Commons were based on exactly the same reasoning, using the same expressions as those employed by Third World countries at the United Nations: cultural dominance, commercial exploitation . . .

To revert to the Clyne Committee, the dangers foreseen, if protective and corrective measures are not quickly devised and implemented, include:

— a reduction in Canadian control over disruption in services resulting from technical breakdowns or work stoppages in another country;
— a reduction in Canadian power to ensure protection against other events, such as invasion of personal privacy or computer crime;
— greater dependence on foreign computer staff which would result in lower requirements for Canadian expertise and a smaller human and technological resource base upon which systems specifically geared to Canadian requirements could be developed;
— jeopardizing the exercise of Canadian jurisdiction over companies operating in Canada which store and process their data abroad;
— undermining the telecommunications system in Canada by the use of foreign communication satellites and roof-top receiving antennas;
— the risk of publication abroad of information that is confidential in Canada.

The Clyne Committee report put forward a number of recommendations for positive action in the form of a strategy to restructure the Canadian telecommunications system so as to make it capable of contributing more effectively to the safeguarding of the sovereignty of the country. The areas that are covered in some depth include the status of the cable industry; the rationalization of the carrier industry; the future of the Canadian Broadcasting Corporation and the private broadcasters; the impact on Canadian social and cultural life of the large-scale importation, via cable, of foreign programming from the television stations in the

United States; a policy for the best use of communications satellites; and the potential threat to Canadian sovereignty posed by informatics and the structure of the electronics manufacturing industry.

A number of the concerns expressed by the Clyne Committee are shared by the report 'L'informatisation de la société', prepared at the request of the French President by Simon Nora and Alain Minc. The similarity of concern is striking between the Clyne Committee and the Nora report, which expresses one of its major themes in the following words: 'If France does not work out the right responses to a whole range of momentous new challenges she will lose the capacity to manage her own affairs. At the heart of this crisis lies the changing nature and importance of information—the "informatization of society"' (or, in another translation: the 'computerization of society').

Like the Clyne Committee, the Nora report emphasizes the need for a national policy. The formation of an informatics policy is an essential measure to prepare for the future. It is urgent to respond rapidly to the challenges of the present which are seen under three main aspects:

— informatics and new economic growth with reference to the risks and opportunities brought about by the new informatics;
— informatics and new power relations which refer to the transformation which informatics will cause in the relations among various economic and social institutions and groups;
— informatics and national independence which has become a new issue of concern through the widening of the areas where sovereignty is important. [See Nora and Minc, 'L'informatisation de la société', 1978.]

Another main theme of relevance in this context is the assertion that the main risks of the information-orientated society are not individual and personal in the form of, for instance, abuses of privacy, but rather the fact that society, as a whole, has become so much more vulnerable and fragile. A modern, industrialized society depends on the efficient working of numerous tiny units. The failure of one part—say, an electricity black-out—can cripple the whole. An information culture that is similarly centralized and hierarchical might push the society beyond the limits of tolerance. The reasons for the present trend towards increased vulnerability do not only derive from the technologies themselves but from the manner in which we have chosen to use them. Although 'chosen' is not the right word, since one of the report's main contentions is that we have arrived at the present situation mainly by default, through a lack of analysis and policy.

The conclusions of the Nora report are strikingly similar to another analysis specifically dealing with the vulnerability of the computerized society. The emphasis of the report by a Swedish Government committee on automated data processing and the vulnerability of society (SÅRK) is that the computerization of society seems inevitable, that the use of computerized systems to a marked degree contributes to the increasing vulnerability of modern, highly industrialized societies, that the level of vulnerability in Sweden is unacceptably high and that it will become even higher if countermeasures are not taken.

The degree of vulnerability is seen as conditioned by a number of factors: the dependence on foreign countries in terms of hardware and software; the centralization and concentration of data systems; the dependence on the few trained operational staff; and the sensitive nature of certain information. Terrorist activities and other criminal actions, threats, sanctions and acts of war are made easier through these vulnerability factors, which also increase the effects of natural catastrophes and accidents.

The Swedish report arrives at many of the same conclusions as the Nora analysis: the lack of awareness of the risks involved, the making of policy by default and the failure to arrive at any conscious national policy. As a final resort, the Swedish Committee even calls for special legislation in the form of a Vulnerability Act to safeguard against the 'unacceptably high risks' in today's computerized society.

The vulnerability created by the large-scale use of interlinked computerized information systems must be put in a wider context. The Swedish Secretariat for Future Studies has undertaken exhaustive studies of the general vulnerability of modern societies as a framework for the specific vulnerabilities arising from new information systems.

Significantly, this report is simply called 'The vulnerable society'. It begins with an attempt to define a frame of reference for the analysis of the vulnerability of modern society, Sweden being used as the main example. The studies included in the report deal with:

(i) psychological and social aspects of vulnerability exemplified by the relationship between confidence in the political system and vulnerability; the normative system as vulnerability surface or a political resource; human judgement and decision-making in a complex society; criminality and the future;

(ii) ecological and technical aspects of vulnerability which deal with the vulnerability of the eco-system, the vulnerability of both small localities and large organizations; data systems and vulnerability.

A major issue concerns the relationship between efficiency and vulnerability. The efforts to achieve ever higher levels of efficiency have a number of implications which increase the level of vulnerability; a series of traditional stabilizing and self-regulating functions has disappeared without being replaced by others; the demands for short-term economic efficiency may put pressure on the environment towards less diversified and therefore more vulnerable eco-systems which can have drastically negative effects on productivity in the long term (the increased use of chemicals in forestry and agriculture can cause a genetic change in the environment through the disappearance of vitally important species).

The report pays particular attention to the problems of size in organizations and the increased use of data systems. With reference to the recent concept 'small is beautiful', the analyses focus on the vulnerability inherent in the growing size of organizations which is seen as a dominant trend in modern society. When a system grows in size there is a corresponding increase in the need for information for control and regulation, i.e. a need for rapid processing of large amounts of information. This trend has resulted in an increased use of computerized information processing. Among the various kinds of vulnerability associated with the large-scale use of computers are mentioned the effects on employment, the dependence on the few experts who design and control the systems, the difficulty of adequately coding information on complex phenomena for computer systems and the inflexibility which is introduced into social organization through computer systems, which make organizations vulnerable to unanticipated changes in their environment.

Concern in Sweden over the vulnerability of an increasingly computerized society led to the establishment, first of a special committee, then of a Vulnerability Board. In a recent statement, the head of the Board emphasized one characteristic of the computerized society: small and simple disturbances may create great and sometimes devastating results. He added a further comment on satellites:

Satellites are being increasingly used for the transmission of information. What will the consequences be when national frontiers are obliterated to such an extent that production processes in one country are controlled from a computer in another country, perhaps on another continent? [Eriksson, 1983, p. 262.]

Interestingly, an American analysis of the Swedish studies and activities from the point of view of applicability to the United States focused on the 'resiliency of the US Information Society', with a warning, however, of the need for further vigilance. A warning that needs to be taken seriously even in the United States where the degree of vulnerability of computerized data systems has recently been proved by groups of teenagers using personal computers to 'break into' major data banks.

The degree of vulnerability in modern society, dependent on the adequate functioning of its communications and information systems, needs to be put into a larger context. Whatever else the reasons, the shooting down of the Korean airliner over Siberia was obviously due to break-downs in communication. Not only were innocent lives lost but informed observers contemplated the risks involved in further faults in the vital information flows. Thus:

current systems are intended to give warning against all forms of attack and to allow control over all sorts of responses. As a consequence, the warning and operational systems are ever more closely interconnected by ever more sophisticated systems of communication, and very small signals can be felt throughout the whole set of systems. [Bundy, 1983, p. 28.]

Security

In all images, analysis and conduct of international affairs the concept of security is a key element. Originally, security in international relations referred to a state of safety from the effects of aggression. A nation that possessed security was considered competent to protect itself through economic, political or military means from foreign aggression. The notion of security is still widely used, referring to military security; thus the term 'collective security' generally refers to military alliances such as NATO, and the Warsaw Pact.

As with so much else in international relations, the concept of security has recently changed and expanded. Scientific knowledge and technological sophistications have gained in importance as elements in dominant positions and security safeguards. Recently, attention has been drawn to the central role of non-weapon technology in the new era of geopolitics conditioned by nuclear weapons and the opening out of outer space. 'Over the last decade, the harnessing of information technology to military tasks has emerged as the principal driving force in the evolution of weapons design and the shaping the strategic balance' (Deudney, 1983, p. 27). With the world-wide networks of military sensing and surveillance, for computerized command, control and communications, the ability to identify targets and to direct nuclear weapons has become planetary in scale, and demands a rethinking of the traditional security policies.

The changing structures and concerns of international relations have pushed both practice and theory beyond politico-military dimensions of security. In 1973 Mahdi Elmandjra, in a study on the United Nations system, declared unequivocally: 'As a purposive function peace is not limited to collective security in the political and military sense. It englobes international economic security, international social security and international cultural security'. (Elmandjra, 1973, p. 319.)

The link between international security and economic

development has been stressed since the 1960s; and this approach paved the way for extending the security concept to the economic field. One of the bases for the demands of developing countries for a new international economic order is the concept of 'economic security'. The Charter of Economic Rights and Duties of States, adopted by the United Nations General Assembly in 1974 explicitly refers to 'collective economic security for development' (preambular paragraph seven). In adopting this Charter, the Assembly stressed that it 'shall constitute an effective instrument towards the establishment of a new system of international economic relations based on equity, sovereign equality and interdependence of the interests of developed and developing countries.' In the Charter itself the emphasis is on co-operation for the purpose of development, as the shared goal and common duty of all states.

The notion of security may be further extended. The international debate and negotiation on environmental issues may well be analysed in terms of efforts to provide 'ecological security'. Since the Stockholm Environment Conference in 1972, the goal of achieving a wider degree of such security has become a permanent feature of the international agenda. The security-promoting measures have included anti-pollution measures to safeguard the marine environment, protection of eco-systems and wildlife and liability for nuclear damage, as well as such more general agreements as the Nordic Environment Convention of 1974 and the Barcelona Agreement on the protection of the Mediterranean of 1978.

A number of the grand international conferences organized by the United Nations in the 1970s have dealt with issues that are well defined by Elmandjra's expression 'international social security'. This is particularly obvious in the case of the conference which led to the establishment of the World Food Programme as the basis for 'food security' for the needy countries. Similarly, international concern over employment and the human habitat can be seen in the light of applying the security concept to other aspects of social life.

That outer space activities have a clear security dimension is quite evident but this enlarged concept of security can also and most usefully be applied in the communications and information field. Once communications *per se* had become an issue in society, at both national and international levels, it was easy to perceive a definite trend towards policies and action designed to establish or increase 'information security', even though the expression itself was rarely used. The strong linkage between traditional security concerns and information issues at the Conferences on Security and Co-operation in Europe (Helsinki, Belgrade, Madrid) was in itself a clear pointer in this direction. Some of the controversies at these conferences may be analysed in terms of the Western countries apparently enjoying a sense of information security while the USSR and its allies have been intent on safeguarding a quite differently perceived information security.

Another example of the quest for 'information security' is represented by the demands of developing countries for a new international information order. The terminology used in this respect seems almost to recall earlier notions of security in the use of such expressions as 'aggression', 'violation', 'imperialism' and 'neo-colonialism' to describe the present practices. The reasons stated by the developing countries in support of the changes which they demand in present systems and practices, explicitly refer to the harmful effects of hardware and software dominance by the industrialized countries and the negative impact of an untrammelled uni-directional flow of technology and information. In the perspective adopted here, the demands by the developing countries can easily be understood as efforts to achieve 'information security', which could be related to the 'cultural security' mentioned by Elmandjra.

To a large extent international public debate on 'information security' has focused on a relatively limited aspect of information, i.e. the journalistic dimension of news and news reporting. A number of analyses have decried the alleged cultural domination exercised in the media field generally, but most of the discussions in UNESCO, the Non-Aligned Conferences and various non-governmental bodies have concerned information in the form of news.

However, other kinds of information flows have also been discussed in terms of 'information security'. In most cases they are related to international data transmissions of various kinds. Particularly interesting in this context are the negotiations in the UN Outer Space Committee on international legal principles to govern remote sensing of the earth by satellite. The fears and misgivings expressed by many countries concern the collection and use of data about their national environment and natural resources, obtained by other countries. If, for example,

transnational companies have better reconnaissance data than developing countries, they are in a position to bargain more effectively for initial exploration privileges. In addition, to the extent that remote-sensing data contributes to a better monitoring of crop developments and improved yield forecasts, the ability to utilize these data, may, for instance, make it possible for users to position themselves effectively in anticipation of market developments, through spot and forward transactions in commodities. [Remote-Sensing Data: Element in National Resource Negotiation, in Transnational Data Report, Vol. VI, No. 5.]

It was soon agreed that the prevention or restriction of space activity, i.e. the collection of data via satellites would have negative effects on international co-operation in which all countries have an interest. The controversy has rather turned around the communications and information aspect, i.e. the treatment, dissemination and use of data. In this respect, a distinction has been made between 'primary data', meaning data acquired by satellite-borne remote sensors as well as pre-processed products derived from these data and 'analyzed information', which means the end-product resulting from the analytical processes performed on the primary data. Thus, some states maintain that permission even for the collection of data must be given by the 'sensed' state, others focus on authorization of the dissemination of analysed information while still others oppose all restrictions concerning the distribution of data and information. These issues thus lend themselves well to an analysis in terms of the balance to be struck between 'information security' and mutually advantageous flows of information.

Information security issues also underlie the discussions on the regulation of another space activity: direct broadcasting from satellites from one country to other countries. In the protracted discussions in the UN Outer Space Committee the controversy has been between a majority of states that wish to ensure agreement before a country undertakes television broadcasting to another country (or countries) and those that have opposed this demand in the name of the free flow of information. Technical advance and the stated intent by countries to use this new technology made agreement necessary—at least in the guise of technical rules within the framework of the ITU.

Interdependence

'We live in an era of interdependence. This vague phrase expresses a poorly understood but widespread feeling that the very nature of world politics is changing.' (Keohane and Nye, 1977, p. 3.) Interdependence has of late become one of those fashionable catchwords that are used to cover a wide range of phenomena. It is used in economics and development theory and has invaded the international relations field. The warning has been sounded that there is a new rhetoric of interdependence. If, earlier, national security dominated international practice and theory, it has now to share its position with interdependence, as rhetoric and symbol and possibly as an analytical tool.

As with any other rhetoric, interdependence can be used for many purposes. In one perspective it is an expression of laziness or obfuscation: if there is nothing else that can be said about a complex international phenomenon, one can always talk about interdependence. In another sense, the rhetoric is used to gain acceptance for a particular policy position. 'We are all engaged in a common enterprise. No nation or group of nations can gain by pushing beyond the limits that sustain world economic growth.' This statement by Henry Kissinger when Secretary of State in a speech to the UN General Assembly has aptly been described as an attempt to limit demands from the Third World countries and influence domestic attitudes rather than to analyse current reality. More dangerous, because more unreal, is to use interdependence as a necessity, almost as a natural law of world development, independent of policy and structure. There is often a stated or, worse, implicit, assumption that since the survival of the human race is threatened by nuclear war or by ecological breakdown, conflicts of interest between states and groups will disappear. Nothing in the present situation warrants this attitude nor will it easily become a feature of tomorrow's world since it implies that everybody will agree on the causes of, and the solutions to, the problems facing us.

In its simplest form interdependence means mutual dependence. In world affairs, interdependence would thus refer to situations or relationships characterized by reciprocal effects among states or other actors. However, this does not say anything about the quality of this mutual dependence, whether its effects are favourable or unfavourable and to whom. There have now been painful discoveries that the advances in science and technology and their applications have opened resource areas or created anti-resources (e.g. pollution) which cause new forms of interdependence and interdependent sources of conflict. In the case of the US/USSR strategic interdependence the costs are high to both parties and to everybody else; one might even press the argument so far as to view war as an ultimate form of interdependence.

Interdependence can also arise without any intentional act by any one partner in a given situation. It may be that the multinational economic system is in a state of disarray, due as much to bureaucratic ineptitude as to any conscious acts. As implied in the discussion of some aspects of vulnerability, interdependence may, as it were, sneak up on countries which one day wake up to find their possibilities of independent action severely limited, without any other country having taken steps towards establishing an unfavourable situation.

Furthermore, some forms of interdependence are conditioned not by man but by nature. Weather knows no national frontiers, nor do radio frequencies. In this case, the laws of nature create a natural interdependence and have forced humans to act accordingly. It is significant that better weather forecasting and an efficient use of the radio frequency spectrum have conditioned some of the most successful forms of international co-operation.

Generally speaking, interdependence is used with reference to relations between countries whether it be in politics, environment, economics or communications. It would, however, be necessary to add further axes of interdependence, some of which refer to phenomena discussed in previous chapters. The most important in this context seem to be the interdependence of national and international politics, or as some analysts express the same thought, the increasing linkages between domestic and international politics.

Even though real situations are more complex and varied than is usually stated, the interdependence of countries seems relatively easy to grasp. Some countries have tried to practise what is called 'linkage politics' in the sense that, say, issues of aid and trade are linked to issues of security or human rights: these attempts seem to raise the degree of complexity to levels which become unmanageable. Less attention has been paid to the complex, often unexpected, interdependence of issues and issue areas. In this context, some examples will be used from the communications and information field. The radio frequency spectrum and the geo-stationary orbit are now perceived as limited natural resources to be shared and allocated according to competing claims not only of countries but also of services: the claims of different services become interlinked and require agreement on priorities which in turn depend on the widely varying socio-economic needs of countries. Television can no longer be seen as merely a medium of entertainment and information—it also has a development dimension, whether positive or negative. Copyright is not just a means to protect and thereby re-munerate literary and artistic creativity—it is also a means of controlling the flow of information and the spread of culture; thus, copyright legislation cannot be divorced from cultural and information policies.

One reason for the lack of recognition of the inter-dependence of issues seems to be that few policy-makers or even experts are prepared to tackle the complex of different, often highly specialized fields involved. Outer space and communications issues represent—similarly to environmental or development questions—issue areas at a new level of complexity. Instead of the global and holistic approach required, governmental, administrative, industrial and academic structures work against boundary

crossing between disciplines, technologies, bureaucracies—
and mental categories.

One attempt to analyse situations and issues of interdependence starts from the observation that the sense of
urgency and importance given an issue and the manner in
which it will be dealt with depends on the kind of interdependence situation that exists. Obviously there is a difference between mutual dependence resulting from nuclear
proliferation or a cut in oil production and the mutual
dependence arising from the need to organize the use of the
radio frequency spectrum on which so many services vital
to modern society depend. Therefore, one analysis, by the
American scholar J. G. Ruggie, points to the structural
factors that distinguish one situation from another. This
analysis is closely linked to the trend towards the increasing
politicization of issues that has been discussed in an earlier
chapter.

The 'collective situation' of a group of countries with
regard to a politicized issue may thus be defined in terms of:

(i) The character of the issue, or the type of interdependence it causes; a good example is the type which is
evident when unilateral action by one state raises un-
certainties for others or makes it impossible for them to
engage in a certain action. This type of policy interdependence in the field of communications is exemplified by such issues as the allocation and use of radio
frequencies and the geo-stationary orbit.

(ii) How closely a particular issue is linked to domestic
policy pursuits, or the locus of interdependence. An
issue may be located outside the domestic policy domain
and affect it only indirectly or, at the other end of the
scale, can directly link domestic policies in one state to
those in others. Both kinds apply to the communications
field, as witnessed by the discussions on issues concerning the international flow of information in the European security conference or in UNESCO. In this scheme
there is one such locus of interdependence that is of
great significance in both outer space and communications: the concept of international or global 'commons'

and the concerns that arise from attempts to define or establish national property rights in these commons. Negotiations over the status of the sea-beds is often taken as the major example but similar conflicts of common international and divergent national interests exist in relation to the exploitation of the moon, the use of the radio frequency spectrum and the geo-stationary satellite orbit.

(iii) The distribution of interdependence is taken to mean the extent to which countries share links of interdependence, the degree to which the situation of a given country is characterized by dependence, dominance or self-sufficiency. Countries will not be equally sensitive or vulnerable to a given type of interdependence, which might also vary with time, often in relation to advances in science and technology.

Another approach to interdependence issues would be to consider what distinguishes them from international issues as traditionally perceived. One crucial new factor:

derives from the many ways in which an ever more dynamic technology and ever growing demands on the world's resources have shrunk the geographic, social, economic and political distances that separate states and vastly multiplied the points at which their needs, interests, ideas, products, organizations and publics overlap. [Rosenau, 1980, p. 41.]

To which should be added the impact of modern communications in the process of shrinking the cushion of space and time that previously separated cultures and societies.

With regard to the characteristics of interdependence issues, Rosenau has singled out four salient ones from all the diverse issues of interdependence, which range from monetary stability to food–population ratios, from the uses of outer space, to water pollution, from oil and energy issues, to the demands for a new international order in the world economy or in international information flows. These charactersitics are:

(i) The large degree to which such issues concern highly complex and mostly technical or natural phenomena. Thus, as pointed out earlier, many forms of interdependence are conditioned not by man but by nature. They require a mastery over physical and biological processes that are not easily understood and, even less, controlled. Other issues, such as the maintenance of monetary stability or the redistribution of wealth, require a comprehension of social, cultural and economic processes that are equally complex and perhaps even more difficult to control. These issues therefore require new kinds of advanced scientific knowledge as tools to deal with the new dimensions of vulnerability and interdependence.

(ii) A second major characteristic of interdependence issues is the large degree to which they involve a great number of non-governmental actors. In fact, these issues often directly concern individuals or groups of citizens. They present a decentralized character which is in sharp contrast to the traditional foreign policy issues that mainly concerned the central foreign ministry bureaucracies: 'The actions of innumerable farmers, for example, are central to the problem of increased food production, just as many pollution issues depend on choices made by vast numbers of producers, energy conservation on millions of consumers, and population growth on tens of millions of potential parents.' (Rosenau, idem, p. 44.)

(iii) The combination of the decentralized nature of these issues and the technical know-how provides the third major characteristic. These two features interact in such a manner as to fragment the governmental decision-making processes through which such issues are considered. This fragmentation of decision-making makes it possible for agencies and sub-agencies to push for separate policies in relation to specific domestic clienteles whose interests they are supposed to guard.

(iv) The fourth characteristic of interdependence issues follows from some of those already mentioned. The management and even more the resolution of dispute is often beyond the power of a single government and does not lend itself to unilateral action. Such issues therefore require multilateral co-operation among governments to an unprecedented degree.

The structure and characteristics of interdependence singled out in these new approaches to international relations provide useful insights into such new issue areas as outer space, information and satellite communication. A prime example is the complex issue area which subsumes such concepts as 'freedom of information' and 'free flow of information', as reflected in the debates on the world information and communications order, direct satellite broadcasting and transborder data flows.

In this issue area international controversy is not only based on the usual defence of national interests: the controversy goes deeper and is more complex. The differences in national interpretations of such principles as freedom of information and their location in a hierarchy of social values, automatically cause different sets of linkages between domestic attitudes, national policy and international politics. The same holds true in informatics with regard to the opposition between 'free data flow' and 'regulated data flow' or in satellite broadcasting with regard to the need for 'prior arrangement'.

These controversies also depend on the difficulties states experience in establishing coherent priorities. The traditional assumption that states act accordingly to relatively well defined, coherent policies following a hierarchically ordered set of issues is particularly unreal in the space and communications field. In fact this assumption presupposes an extraordinary degree of consensus at the national level, and of co-ordination between national bureaucracies and an absence of conflicts of interest even on controversial issues. A case might be made that such is still the situation in some countries on some issues. It is certainly contrary to observed behaviour on most communications issues that are under international debate. On the contrary, this area is one of great confusion and rather presents an extreme example of a common feature in most new thinking on international relations: a multiplicity of issues which are given very varied orders of priority, thus a lack of hierarchy of issues.

The lack of an ordered set of policies does not only concern issues within the communications field but also the relative priority assigned to communications in relation to other isssue areas. One crucial issue is related to the already mentioned extension of the security concept and can be exemplified by the different weight given by countries to the principle of freedom of information relative to the principle of non-interference in domestic affairs. The discussions at the Helsinki, Belgrade and Madrid conferences on security and co-operation in Europe have clearly shown the different priorities given to 'information security'.

The lack of a clear hierarchy of issues is interesting also in that it has been used as a major element in a recent model or paradigm in international relations known as 'complex interdependence'. As proposed by the American scholars Keohane and Nye, it is characterized by three main features:

(i) Multiple channels connecting societies. These channels include not only the formal arrangements between foreign offices but also the informal ties between governmental and non-governmental elites as well as transnational organizations (multinational companies, etc.).

(ii) The state of complex interdependence also implies that the agenda of inter-state relationships includes multiple issues that are not arranged in a clear, consistent hierarchy. This absence of a clear hierarchy or order of priority implies that any one issue—say, military security— does not consistently dominate the agenda. Issues are considered by several government departments and at various levels. Different issues therefore generate different coalitions, both within and across governments and involve different degrees of co-operation or conflict.

(iii) Military force as an instrument of policy is not used by governments towards other governments within the region, or on the issues where complex interdependence prevails.

This model well describes certain characteristic features of international relations in the information/communications field. A key feature is the phenomenon of multiple channels, even beyond what the originators of this model have stated. Informal ties, at least between some societies, do not only involve traditional elites but a wide range of groups and interests which establish cross-national links. In the area of communications, the interstate agenda shows consistency of priorities only to a limited and often unpredictable extent since priorities change both with issues and with forums. So far, military force has not been an instrument of policy in communications issues but the militarization of space might lead to a different situation.

These recent paradigms in the analysis of international relations and world politics, therefore, provide useful insights and new intellectual tools in the consideration of such new and complex international issue areas as outer space, information—and satellite communications.

New intellectual tools will certainly be required to implement the conclusions and recommendations of UNISPACE 82, the Second United Nations Conference on the Exploration of Peaceful Uses of Outer Space which was held in Vienna in August 1982. The Conference provided an opportunity for the international community to consider in detail a great number of the complex issues raised by the rapid progress in the development of space technology, and particularly the vast increase in actual and potential applications. The developing countries which had initially taken much of the initiative for convening UNISPACE 82 expressed the desire to explore how the world-wide activities in outer space would be developed to ensure that the potential benefits from space science and technology would be truly realized for all countries. There seems to have been a general feeling that the potential for space was much greater than is currently being realized by most countries but that the benefits were not shared as widely as they could be.

The Conference dealt with the entire gamut of space sciences, technologies and applications from the scientific, technical, political, economic, social and organizational points of view. Thus, the report of the Conference deals with matters relating to the prevention of an arms race in space, the need and possibilities for technology transfer, co-ordination in the use of the geo-stationary orbit, remote sensing of earth resources from space, the monitoring of pollution and changes in atmospheric conditions, the use of direct broadcast satellites, particularly as an aid to the spread of education, space transportation and space platform technologies, protection of the near-earth environment, the future operational applications of manufacturing and processing in space, space power systems, the long-term consequences of the increasing number of launchings, and the encouragement of the development of indigenous capacity by developing countries. The recommendations of the Conference which were adopted by the UN General Assembly have even been hailed as an agenda for nations and organizations to follow in the next few decades.

In the autumn of 1982 another agenda for the future was also set, this one by the International Telecommunication Union at its Plenipotentiary Conference held in Nairobi. This agenda appears to be of a more technical nature in the description of the subjects that are to be discussed at various radio administrative conferences during the next ten years or so. However, activities such as a regional plan for the Americas for the allocation of geo-stationary orbit locations and of satellite beams for direct satellite broadcasting or a review of the allocations and uses of the high-frequency band have great economic and political implications. Even more important will be the issues and the implications of the World Administrative Radio Conference starting in 1985 which will review the technical, administrative and regulatory aspects of all space services. Still another agenda has been adopted by the UNESCO General Conference in the form of a Medium-Term Plan for the period 1984–9 in which the programmes envisaged in the field of communications and information would be of relevance in this context.

And these agendas for the future development of outer space activities, and of communications and information represent only the tip of an iceberg. Other international and regional organizations have their own outlook, priorities and preoccupations, to which should be added the agendas, open or hidden, set by major corporations, by national governments, by professional organizations and interest groups, and by laboratories and research institutions. All will influence the overall agenda of our planet.

10 Epilogue

Many agendas are being set for the planet. Assuming that we can ward off the worst, whether in the form of a nuclear catastrophe or an intolerable erosion of our physical and social environment, I should like, in the context of this book, to suggest three specific agenda items.

The first item for this planetary agenda is to safeguard the global commons of our earth or—in another version—the common heritage of mankind. In fact, it is a heritage which mankind shares with other living beings and with all elements that make up our common environment: rocks and trees, climate and weather, land, sea, air—and outer space. There is more though. Just as we need to safeguard the genetic diversity of our physical environment, we also need to safeguard the cultural diversity of our social environment. We require both: the disappearance of a species might foreclose openings in the future, as may the disappearance of cultural inventions. Knowledge and information are resource realms as precious as any other, perhaps more so in the sense that they condition our attitudes to other resources in our environment, the manner in which we use or misuse them. The safeguarding of these global commons means solidarity —solidarity in the present to encompass also outer space, as our new fourth environment; solidarity also in time, an inter-generational solidarity. In this perspective, a proposal of principles to guide us in our management of natural resources is equally valid for our management of cultural resources: conservation of options, which amounts to conserving diversity in the resource base; and conservation of quality, defined as leaving the planet in no worse condition than the one in which it was received. (See Brown-Weiss, 1983.)

The second item for the planetary agenda follows directly from the first one: the need for learning how to manage complexity, how to deal with global problems and relate them to problems at the local, national and regional level, how to guide social transformation and adjust to continuous and often painful change, how to reach for solutions to the pressing problems of poverty, hunger and resource use, to the arms race and the disarray of the global economy, to the generation, dissemination and application of knowledge. We need to manage complex physical and social systems to deal with new levels of complexity caused by the interdependence of countries and of issues.

A hopeful sign is that work in specialized sciences and convergence of findings about the characteristics and behaviour of complex systems are resulting in the emergence of a new science of complexity. Relevant findings and new ideas are to be found in such classical disciplines as physics, chemistry, biology, neurophysiology and economics, as well as in more recent branches of knowledge such as general systems theory, telecommunications, computer and information science, ecology and environmental sciences, or in new approaches developed within traditional disciplines such as geography, urban studies, cognitive sciences. The study of complexity is located at the frontiers of knowledge. New scientific paradigms are emerging, more receptive to a plurality of cultural traditions, seen even as representing a metamorphosis of science, an 'opening-up of a new theoretical space'. (Ilya Prigogine.)

The third item concerns the need to increase our learning capacity. The concept of learning as used here, and earlier in the book, goes beyond what is implied in traditional terms such as education, teaching or schooling. Learning is mediated by the total environment, physical and social. It signifies an approach to life and to knowledge which is active, participatory and anticipatory, listening as much as teaching. This, in turn, implies the need for developing new, individual as well as institutional and societal learning capacities, the need for improving social preparedness and the capacity to cope with the new levels of complexity reflected in the growing interdependence of societies and of issues.

The capacity of a nation—not just of its government, but of society as a whole—to adjust to rapidly changing techno-economic, sociocultural and political changes, on a scale which makes it possible to speak of social transformation, very much depends on its collective capacity to generate, to ingest, to reach out for, and to utilize a vast amount of new and relevant information. This capacity for creative and innovative response to changing conditions and new challenges I should like to call the learning capacity of a nation. This capacity is obviously not limited to the cognitive level, but includes the attitudinal, institutional and organizational levels of society as well. [Soedjatmoko, 1978, p. 15.]

The spirit in which to safeguard our common heritage and to deal with complexity, to increase our learning capacity, has been strikingly expressed by the late Barbara Ward: the need to be loyal to all mankind—and to our earth,

alone in space, alone in its life-supporting systems, powered by inconceivable energies, wayward, unlikely, unpredictable, but nourishing, enlivening and enriching in the largest degree—is this not a precious home for all of us earthlings? Is it not worth our love? Does it not deserve all the inventiveness and courage and generosity of which we are capable to preserve it from degradation and by doing so, to secure our own survivial?

Bibliographic References

Branscomb, Anne W., 'Toward a Global Communications Policy for an Interdependent World', unpublished, 1980.

Brewin, Bob, 'The New Satellite Wave', *Soho News*, 5 January 1982.

Brown-Weiss, Edith, 'Principles for Resolving Conflicts between Generations over New Natural Resources', *Mazingira*, Vol. 7, No. 2, 1983.

Bundy, McGeorge, 'Ending War Before It Starts', *New York Times Book Review*, 9 October 1983.

Butler, Richard, presentation of the ITU/OECD Synthesis Report, 'Telecommunications for Development', World Communications Year Seminar/Meeting, San José, 8 August 1983.

Calder, Nigel, 'Our Small and Lovely Planet', *New Statesman*, 3 January 1969.

Calder, Nigel, *Violent Universe*, London, Future Publications, 1975.

Chapman, John H. and Warren, Gabriel, 'Direct Broadcast Satellites: the ITU, UN and the Real World', *McGill Annals*, 1979.

Clarke, Arthur C., *Voices from the Sky*, New York, Harper & Row, 1967.

Clarke, Arthur C., ed., *The Coming of the Space Age*, London, Panther Books, 1970.

Consultative Committee on the Implications of Telecommunications for Canadian Sovereignty, *Telecommunications and Canada*, Ottowa, Ministry of Supply and Services, 1979.

Council of Europe, Document 2051, 20 April 1966.

Devieux, Charles, 'France Moves towards New Services', *InterMedia*, July 1981.

Deudney, Daniel, *Space: The High Frontier in Perspective*, Worldwatch Paper No. 50, Worldwatch Institute, Washington DC, 1982.

Deudney, Daniel, 'Whole Earth Security: A Geopolitics of Peace', Worldwatch Paper No. 55, Washington DC, Worldwatch Institute, 1983.

Deutsch, Karl W., 'Outer Space and International Politics: A Look to 1988', in Goldsen, Joseph M., ed., *Outer Space in World Politics*, London, Pall Mall Press, 1963.

Ehrecke, Krafft, 'The Anthropology of Space Flight', in Clarke, Arthur C., ed., *The Coming of the Space Age*, London, Panther Books, 1970.

Elmandjra, Mahdi, *The United Nations System: an Analysis*, London, Faber and Faber, 1973.

Eriksson, Allan, 'Report to the Securicom—83 Conference', Cannes, published in *Transnational Data Report*, Vol. 6, No. 5, July/August 1983.

Fox, Barry, 'Satellite TV Starts the Ultimate Craze', *New Scientist*, 9 September 1982.

Friedman, Robert, 'Supernews', in *Channels*, September/October, 1983.

Ganley, Oswald H., 'The International Arena: Some Preliminary Thoughts', unpublished, 1980.

Golden, David A., 'Technology—Master or Servant', The 1982 Louis G. Cowan Lecture, *InterMedia*, Vol. 10, No. 6, November 1982.

Goldsen, Joseph M., ed., *Outer Space in World Politics*, London, Pall Mall Press, 1963.

Goulden, J., *Monopoly*, New York, Putnam, 1968.

Halloran, Richard, 'US Plans Big Spending Increase for Military Operations in Space', *New York Times*, 17 October 1982.

Hayes, E. Nelson, 'Tracking Sputnik I', in Clarke, Arthur C., ed., *The Coming of the Space Age*, London, Panther Books, 1970.

Hondius, Fritz, 'Reducing Media Turmoil in Europe', *Transnational Data Report*, Vol. 4, No. 3, April/May 1983.

Howard, Mike, 'Geo-stationary Environmental Satellites', *Spaceflight*, Vol. 16, No. 10, October 1974.

Hoyle, Fred, *The New Face of Science*, New York, The New American Library Inc., 1971.

Hoyle, Fred, *Astronomy Today*, London, Heinemann Educational Books Ltd, 1975.

International Institute of Communications, 'Telecommunications— National Policy and International Agreement', A Briefing Paper in preparation for the World Administrative Radio Conference of 1979, London, 1977.

Jasani, Bhupandra, ed., *Outer Space—A New Dimension of the Arms Race*, London, Stockholm International Peace Research Institute/ Taylor & Francis Ltd, 1982.

Johnson, Nicholas, 'The Media Barons and the Public Interest', *The Atlantic*, May 1968.

Karnik, Kiran, ed., *Alternative Space Futures and the Human Condition*, Oxford, published for the United Nations by Pergamon Press, 1982.

Kenden, Anthony, 'U.S. Reconnaisance Satellite Programmes', *Spaceflight*, Vol. 20, No. 7, July 1978.

Keohane, Robert and Nye, Joseph, *Power and Interdependence*, Boston, Little, Brown & Company, 1977.

Krishnamoorthy, P. V., 'A Satellite in the Service of the Underprivileged: The Indian Experiment with SITE', *EBU Review*, May 1977.

Lindgren, Kerstin, 'Sweden Turns from Nordsat to its own Tele-X', *InterMedia*, July 1981.

McElroy, John H., 'Telecommunications and the Global View', paper prepared for the 1983 Annual Conference of the International Institute of Communications, unpublished.

Mohr, Charles, 'It's a Long Way to Star Wars', *New York Times*, 27 June 1982, p. E9.

Müller, Albrecht, 'Commercial satellites may threaten West Germany's media mix', *InterMedia*, July 1981.

Nora, Simon and Minc, Alain, 'L'informatisation de la société', Paris, *Le documentation française*, 1978.

Oberg, James, 'Crisis in orbit for the USSR', *New Scientist*, 24 November 1983.

Oberg, James, 'Russians in Orbit', *Future Life*, No. 30, November 1981.

Pal, Yash, 'Prospects and Challenges in Communication', unpublished 3 August, 1983.

Pal, Yash, 'Some Thoughts on Remote Sensing', paper prepared for the 1983 Annual Conference of the International Institute of Communications, unpublished.

Pelton, Joseph, Perras, Marcel, and Sinha, Ashok K., 'Intelsat, the Global Telecommunications Network', document presented at Pacific Telecommunications Conference, Honolulu, Hawaii, January 1983.

Petrov, Boris, 'Russia's Space Future', *Spaceflight*, Vol. 16, No. 11, November 1974.

Ploman, Edward W., *Kommunikation durch Satelliten*, Mainz, v. Hase & Koehler Verlag, 1974.

Ploman, Edward W., 'Satellite Broadcasting: Promise and Reality', *EBU Review*, Vol. XXVIII, No. 3, May 1977.

Ploman, Edward W. and Hamilton, L. Clark, *Copyright: Intellectual Property in the Information Age*, London, Routledge & Kegan Paul, 1980.

Ploman, Edward W., *International Law Governing Communications and Information*, London, Frances Pinter, 1982.

Possony, Stefan T. and Rosenzweig, Leslie, 'The Geography of the Air, 1955', quoted in Donald Cox and Michael Stoiki, *Spacepower*, Philadelphia, The John C. Winston Company, 1958.

Rahim, Syed A., 'Telecommunications Technology and Policy: The Case of Satellite Communication', *Keio Communication Review*, No. 4, March 1983.

Rosenau, James N., *The Study of Global Interdependence: Essays on the Transnationalization of World Affairs*, London, Frances Pinter, New York, Nichols Publishing Company, 1980.

Ruggie, John Gerard, 'International Responses to Technology: Concepts and Trends, *International Organization*, Vol. 29, No. 3, 1975.

Rutkowski, Anthony M., 'ISDN: Designing the World's Telecommunications Networks', *InterMedia*, March 1983, Vol. 2, No. 2.

Schwarz, Michiel, 'The Politics of European Space Collaboration', *InterMedia*, Vol. 9, No. 4, July 1981.

Rutkowski, Anthony M., 'ISDN: Designing the World's Telecommunications Networks', *Intermedia*, March, 1983, Vol. 2, No. 2.

Schwarz, Michiel, 'The Politics of European Space Collaboration', *Intermedia*, Vol. 9, No. 4, July 1981.

Siuru, W. D. and Holder, William G., 'Earth Resources Satellites: Rescuing Earth from Outer Space', *Spaceflight*, Vol. 14, No. 2, May 1972.

Smith, Delbert, D., *Space Stations, International Law and Policy*, Boulder, Colorado, Westview Press, 1979.

Smith, Delbert, D., *Communications via Satellite*, Leyden, A. W. Sijthoff/ Boston, Mass., A. W. Sijthoff, 1976.

Soedjatmoko, *The Future and the Learning Capacity of Nations: The Role of Communications*, London, International Institute of Communications, 1978.

Stapledon, Olaf, 'Interplanetary Man', in Clarke, Arthur C., ed., *The Coming of the Space Age*, London, Panther Books, 1970.

Stine, Harry G., 'The Third Industrial Revolution: The Exploitation of the Space Environment', *Spaceflight*, Vol. 16, No. 9, September 1974.

Supandhiloke, Boonlert, 'Satellite Communications Programs in Thailand', *Keo Communication Review*, No. 4, March 1983.

Svenson, Ola Och Wentzel, Ann-Kristin, 'Det Sårbara Samhället', Stockholm, Sekreteriatet För Framtidsstudier, 1978.

Sårbarhetskommittén (SÅRK), 'ABD och Samhällets Sårbarhet', Stockholm, Försvarsdepartementet, 1978.

United Nations, 'Current and Future State of Space Science, Background Paper, Second United Nations Conference on the Exploration and Peaceful Uses of Outer Space', Document A/CONF. 101/BP/1.

United Nations, 'Feasibility and Planning of Instructional Satellites Systems, Background Paper, Second United Nations Conference on the Exploration and Peaceful Uses of Outer Space', Document A/CONF. 101/BP/6.

United Nations, 'Relevance of Space Activities to Monitoring of Earth Resources and the Environment, Background Paper, Second United Nations Conferfence on the Exploration and Peaceful Uses of Outer Space', Document A/CONF. 101/BP/3.

United Nations, 'Report of the Second United Nations Conference on the Exploration and Peaceful Uses of Outer Space', Vienna, 9–12 August 1982, Document A/CONF. 101/10.

United Nations Development Programme, 'Development Information Network for Co-operation among Developing Countries', New York, 1980.

United Nations Economic Commission for Africa, 'PADIS: Pan African Documentation and Information System', Document PADIS/DEVSIS—Africa/80/6, Addis Ababa, 1980.

United Nations Economic Commission for Africa, 'PADIS, Pan African Documentation and Information System', Addis Ababa, 1981.

United Nations Educational, Scientific and Cultural Organization (UNESCO), 'Global Satellite Project for Dissemination and Exchange of Information: Report on Mission to Washington and Meeting with Intelsat', Document CONF/FCP/3'78, 15 January 1982.

United Nations Educational, Scientific and Cultural Organization (UNESCO), 'The Use of Satellite Communication for Information Transfer, prepared by the International Institute of Communications for the General Information Programme and UNISIST', Paris, 1982.

Waldrop, Mitchell M., 'Imaging the Earth (II): The Politics of Landsat', *Science*, Vol. 216, 2 April 1982.

Wilford, John Noble, 'Space Telescope Holds NASA's Hopes for Grand Discoveries in the Universe', *New York Times*, 5 January 1982.

Index